McGraw Hill
Illustrative Mathematics®
Algebra 1

Cover Credit: Anna Bliokh/iStockphoto/Getty Images

mheducation.com/prek-12

Illustrative Math Algebra 1, Geometry, and Algebra 2 are ©2019 Illustrative Mathematics.
Modifications ©McGraw Hill.

Send all inquiries to:
McGraw Hill
8787 Orion Place
Columbus, OH 43240

ISBN: 978-0-07-693052-4
MHID: 0-07-693052-1

Illustrative Mathematics, Algebra 1
Student Edition, Volume 2

Printed in the United States of America.

10 11 12 13 14 15 16 MER 28 27 26 25 24 23

'Notice and Wonder' and 'I Notice/I Wonder' are trademarks of the National Council of Teachers of Mathematics, reflecting approaches developed by the Math Forum (http://www.nctm.org/mathforum/), and used here with permission.

Contents in Brief

Welcome to

McGraw Hill
Illustrative
Mathematics®

Illustrative Mathematics is designed to provide opportunities for you to learn math by doing math. By engaging in problem-solving activities, you will make and test your own conclusions and try out different strategies as you learn math.

Let's get started!

Unit 1

One-variable Statistics

Unit 2

Linear Equations, Inequalities, and Systems

Systems of Linear Inequalities in Two Variables

Unit 3
Two-variable Statistics

Unit 4

Functions

Unit 5

Introduction to Exponential Functions

Unit 6

Introduction to Quadratic Functions

Unit 7

Quadratic Equations

Vertex Form Revisited

Putting It All Together

Introduction to Exponential Functions

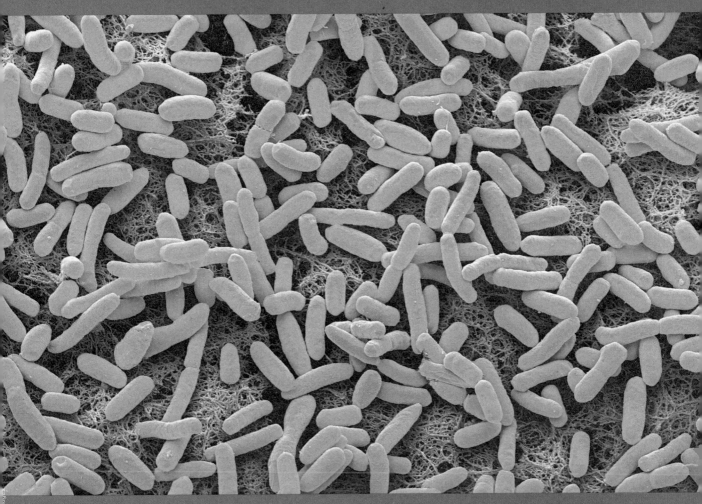

An exponential function can be used to model the growth of bacteria.
You will learn more about exponential functions in this unit.

Topics
- Looking at Growth
- A New Kind of Relationship
- Exponential Functions
- Percent Growth and Decay
- Comparing Linear and Exponential Functions
- Putting It All Together

Introduction to Exponential Functions

Lesson 5-1

Growing and Growing

NAME _____ DATE _____ PERIOD _____

Learning Goal Let's choose the better deal.

 Warm Up
1.1 Splitting Bacteria

There are some bacteria in a dish. Every hour, each bacterium splits
into 3 bacteria.

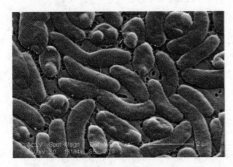

1. This diagram shows a bacterium in hour 0 and then hour 1. Draw what
   happens in hours 2 and 3.

2. How many bacteria are there in hours 2 and 3?

Activity

1.2 A Genie in a Bottle

You are walking along a beach and your toe hits something hard. You reach down, grab onto a handle, and pull out a lamp! It is sandy. You start to brush it off with your towel. Poof! A genie appears.

He tells you, "Thank you for freeing me from that bottle! I was getting claustrophobic. You can choose one of these purses as a reward."

- Purse A which contains $1,000 today. If you leave it alone, it will contain $1,200 tomorrow (by magic). The next day, it will have $1,400. This pattern of $200 additional dollars per day will continue.

- Purse B which contains 1 penny today. Leave that penny in there, because tomorrow it will (magically) turn into 2 pennies. The next day, there will be 4 pennies. The amount in the purse will continue to double each day.

1. How much money will be in each purse after a week? After two weeks?

2. The genie later added that he will let the money in each purse grow for three weeks. How much money will be in each purse then?

3. Which purse contains more money after 30 days?

NAME _____ DATE _____ PERIOD _____

Activity
1.3 Graphing the Genie's Offer

Here are graphs showing how the amount of money in the purses changes. Remember Purse A starts with $1,000 and grows by $200 each day. Purse B starts with $0.01 and doubles each day.

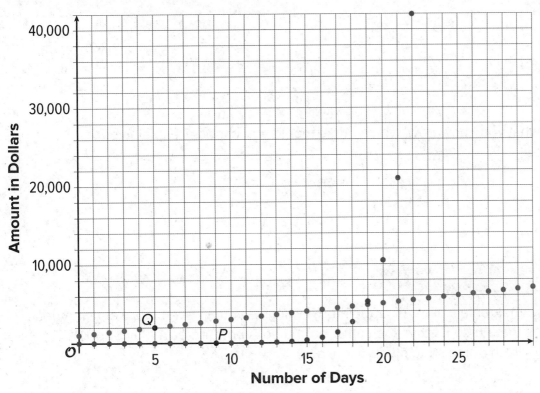

1. Which graph shows the amount of money in Purse A? Which graph shows the amount of money in Purse B? Explain how you know.

2. Points P and Q are labeled on the graph. Explain what they mean in terms of the genie's offer.

3. What are the coordinates of the vertical intercept for each graph? Explain how you know.

4. When does Purse B become a better choice than Purse A? Explain your reasoning.

5. Knowing what you know now, which purse would you choose? Explain your reasoning.

Are you ready for more?

Okay, okay, the genie smiles, disappointed. I will give you an even *more* enticing deal. He explains that Purse B stays the same, but Purse A now increases by $250,000 every day. Which purse should you choose?

NAME _____ DATE _____ PERIOD _____

Summary
Growing and Growing

When we repeatedly double a positive number, it eventually becomes *very* large. Let's start with 0.001. The table shows what happens when we begin to double:

0.001	0.002	0.004	0.008	0.016

If we want to continue this process, it is convenient to use an exponent. For example, the last entry in the table, 0.016, is 0.001 being doubled 4 times, or $(0.001) \cdot 2 \cdot 2 \cdot 2 \cdot 2$, which can be expressed as $(0.001) \cdot 2^4$.

Even though we started with a very small number, 0.001, we don't have to double it that many times to reach a very large number. For example, if we double it 30 times, represented by $(0.001) \cdot 2^{30}$, the result is greater than 1,000,000.

Throughout this unit, we will look at many situations where quantities grow or decrease by applying the same factor repeatedly.

Practice
Growing and Growing

1. Which expression equals 2^7?

   (A.) $2 + 2 + 2 + 2 + 2 + 2 + 2$ (C.) $2 \cdot 7$

   (B.) $2 \cdot 2 \cdot 2 \cdot 2 \cdot 2 \cdot 2 \cdot 2$ (D.) $2 + 7$

2. Evaluate the expression $3 \cdot 5^x$ when x is 2.

3. The graph shows the yearly balance, in dollars, in an investment account.

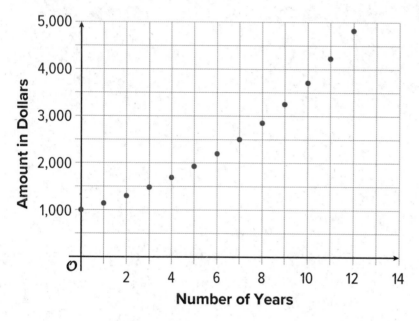

   a. What is the initial balance in the account?

   b. Is the account growing by the same number of dollars each year? Explain how you know.

   c. A second investment account starts with $2,000 and grows by $150 each year. Sketch the values of this account on the graph.

   d. How does the growth of balances in the two account balances compare?

NAME _____ DATE _____ PERIOD _____

4. Jada rewrites $5 \cdot 3^x$ as $15x$. Do you agree with Jada that these are equivalent expressions? Explain your reasoning.

5. Investment account 1 starts with a balance of $200 and doubles every year. Investment account 2 starts with $1,000 and increases by $100 each year.

   a. How long does it take for each account to double?

   b. How long does it take for each account to double again?

   c. How does the growth in these two accounts compare? Explain your reasoning.

6. A study of 100 recent high school graduates investigates a link between their childhood reading habits and achievement in high school.

   Participants are asked if they read books every night with another person when they were ages 2 to 5, as well as their grade average for all of their high school classes. The results are represented in the table. (Lesson 3-1)

	Read Books Nightly	Did Not Read Books Nightly
A average	16	10
B average	21	14
C average	12	16
D average	3	8

   a. What does the 21 in the table represent?

   b. What does the 10 in the table represent?

7. Lin says that a snack machine is like a function because it outputs an item for each code input. Explain why Lin is correct. (Lesson 4-1)

8. At a gas station, a gallon of gasoline costs $3.50. The relationship between the dollar cost of gasoline and the gallons purchased can be described with a function. (Lesson 4-1)

   a. Identify the input variable and the output variable in this function.

   b. Describe the function with a sentence of the form "_____ is a function of _____."

   c. Identify an input-output pair of the function and explain its meaning in this situation.

Lesson 5-2

Patterns of Growth

NAME _____ DATE _____ PERIOD _____

Learning Goal Let's compare different patterns of growth.

 ## Warm Up
2.1 Which One Doesn't Belong: Tables of Values

Which one doesn't belong?

Table A

x	y
1	8
2	16
3	24
4	32
8	64

Table B

x	y
0	0
2	16
4	32
6	48
8	64

Table C

x	y
0	1
1	4
2	16
3	64
4	256

Table D

x	y
0	4
1	8
2	12
3	16
4	20

Activity

2.2 Growing Stores

A food company currently has 5 convenience stores. It is considering 2 plans for expanding its chain of stores.

Plan A: Open 20 new stores each year.

1. Use technology to complete a table for the number of stores for the next 10 years, as shown here.

Year	Number of Stores	Difference From Previous Year
0	5	
1	25	
2		
3		
4		
5		
6		
7		
8		
9		
10		

2. Respond to each question.

   a. What do you notice about the difference from year to year?

   b. If there are n stores one year, how many stores will there be a year later?

3. Respond to each question.

   a. What do you notice about the difference every 3 years?

   b. If there are n stores one year, how many stores will there be 3 years later?

NAME _____ DATE _____ PERIOD _____

Plan B: Double the number of stores each year.

1. Use a technology to complete a table for the number of stores for the next 10 years under each plan, as shown here.

Year	Number of Stores	Difference From Previous Year	Factor From Previous Year
0	5		
1			
2			
3			
4			
5			
6			
7			
8			
9			
10			

2. Respond to each question.

   a. What do you notice about the difference from year to year?

   b. What do you notice about the factor from year to year?

   c. If there are n stores one year, how many stores will there be a year later?

3. Respond to each question.

   a. What do you notice about the difference every 3 years?

   b. What do you notice about the factor every 3 years?

   c. If there are n stores one year, how many stores will there be 3 years later?

Suppose the food company decides it would like to grow from the 5 stores it has now so that it will have at least 600 stores, but no more than 800 stores 5 years from now.

1. Come up with a plan for the company to achieve this where it adds the same number of stores each year.

2. Come up with a plan for the company to achieve this where the number of stores multiplies by the same factor each year. (Note that you might need to round the outcome to the nearest whole store.)

Activity

2.3 Flow and Followers

Here are verbal descriptions of 2 situations, followed by tables and expressions that could help to answer one of the questions in the situations.

- Situation 1: A person has 80 followers on social media. The number of followers triples each year. How many followers will she have after 4 years?

- Situation 2: A tank contains 80 gallons of water and is getting filled at rate of 3 gallons per minute. How many gallons of water will be in the tank after 4 minutes?

Match each representation (a table or an expression) with one situation. Be prepared to explain how the table or expression answers the question.

A. $80 \cdot 3 \cdot 3 \cdot 3 \cdot 3$

D. $80 + 4 \cdot 3$

B.

x	0	1	2	3	4
y	80	240	720	2,160	6,480

E.

x	0	1	2	3	4
y	80	83	86	89	92

C. $80 + 3 + 3 + 3 + 3$

F. $80 \cdot 81$

NAME _____ DATE _____ PERIOD _____

Summary
Patterns of Growth

Here are two tables representing two different situations.

- A student runs errands for a neighbor every week. The table shows the pay he may receive, in dollars, in any given week.

Number of Errands	Pay in Dollars	Difference From Previous Pay
0	10	
1	15	5
2	20	5
3	25	5
4	30	5

- A student at a high school heard a rumor that a celebrity will be speaking at graduation. The table shows how the rumor is spreading over time, in days.

Day	People Who Have Heard the Rumor	Factor From Previous Number of People
0	1	
1	5	5
2	25	5
3	125	5
4	625	5

Once we recognize how these patterns change, we can describe them mathematically. This allows us to understand their behavior, extend the patterns, and make predictions.

In upcoming lessons, we will continue to describe and represent these patterns and use them to solve problems.

Practice
Patterns of Growth

1. A population of ants is 10,000 at the start of April. Since then, it triples each month.

   a. Complete the table.

   b. What do you notice about the population differences from month to month?

Months Since April	Number of Ants
0	
1	
2	
3	
4	

   c. If there are n ants one month, how many ants will there be a month later?

2. A swimming pool contains 500 gallons of water. A hose is turned on, and it fills the pool at a rate of 24 gallons per minute. Which expression represents the amount of water in the pool, in gallons, after 8 minutes?

   A. $500 \cdot 24 \cdot 8$

   B. $500 + 24 + 8$

   C. $500 + 24 \cdot 8$

   D. $500 \cdot 24^8$

3. The population of a city is 100,000. It doubles each decade for 5 decades. Select **all** expressions that represent the population of the city after 5 decades.

   A. 32,000

   B. 320,000

   C. $100,000 \cdot 2 \cdot 2 \cdot 2 \cdot 2 \cdot 2$

   D. $100,000 \cdot 5^2$

   E. $100,000 \cdot 2^5$

NAME _____ DATE _____ PERIOD _____

4. The table shows the height, in centimeters, of the water in a swimming pool at different times since the pool started to be filled.

   a. Does the height of the water increase by the same amount each minute? Explain how you know.

Minutes	Height
0	150
1	150.5
2	151
3	151.5

   b. Does the height of the water increase by the same factor each minute? Explain how you know.

5. Bank account C starts with $10 and doubles each week. Bank account D starts with $1,000 and grows by $500 each week.

   When will account C contain more money than account D? Explain your reasoning. (Lesson 5-1)

6. Suppose C is a rule that takes time as the input and gives your class on Monday as the output. For example, C(10:15) = Biology. (Lesson 4-2)

   a. Write three sample input-output pairs for C.

   b. Does each input to C have exactly one output? Explain how you know.

   c. Explain why C is a function.

7. The rule that defines function f is $f(x) = x^2 + 1$. Complete the table. Then, sketch a graph of function. (Lesson 4-4)

x	f(x)
-4	17
-2	
0	
2	
4	

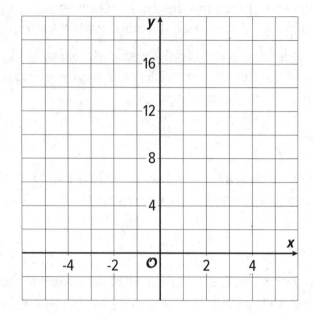

8. The scatter plot shows the rent prices for apartments in a large city over ten years. (Lesson 3-4)

a. The best fit line is given by the equation $y = 134.02x + 655.40$, where y represents the rent price in dollars, and x the time in years. Use it to estimate the rent price after 8 years. Show your reasoning.

b. Use the best fit line to estimate the number of years it will take the rent price to equal $2,500. Show your reasoning.

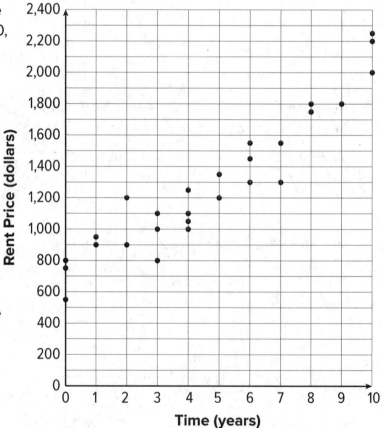

Lesson 5-3

Representing Exponential Growth

NAME _____ DATE _____ PERIOD _____

Learning Goal Let's explore exponential growth.

Warm Up
3.1 Math Talk: Exponent Rules

Rewrite each expression as a power of 2.

$2^3 \cdot 2^4$

$2^5 \cdot 2$

$2^{10} \div 2^7$

$2^9 \div 2$

Activity

3.2 What Does x^0 Mean?

1. Complete the table. Take advantage of any patterns you notice.

x	4	3	2	1	0
3^x	81	27			

2. Here are some equations. Find the solution to each equation using what you know about exponent rules. Be prepared to explain your reasoning.

   a. $9^? \cdot 9^7 = 9^7$

   b. $\dfrac{9^{12}}{9^?} = 9^{12}$

3. What is the value of 5^0? What about 2^0?

Are you ready for more?

We know, for example, that $(2 + 3) + 5 = 2 + (3 + 5)$ and $2 \cdot (3 \cdot 5) = (2 \cdot 3) \cdot 5$. The grouping with parentheses does not affect the value of the expression.

Is this true for exponents? That is, are the numbers $2^{(3^5)}$ and $(2^3)^5$ equal? If not, which is bigger? Which of the two would you choose as the meaning of the expression 2^{3^5} written without parentheses?

NAME _____ DATE _____ PERIOD _____

Activity

3.3 Multiplying Microbes

1. In a biology lab, 500 bacteria reproduce by splitting. Every hour, on the hour, each bacterium splits into two bacteria.

   a. Write an expression to show how to find the number of bacteria after each hour listed in the table.

Hour	Number of Bacteria
0	500
1	
2	
3	
6	
t	

   b. Write an equation relating n, the number of bacteria, to t, the number of hours.

   c. Use your equation to find n when t is 0. What does this value of n mean in this situation?

2. In a different biology lab, a population of single-cell parasites also reproduces hourly. An equation which gives the number of parasites, p, after t hours is $p = 100 \cdot 3^t$. Explain what the numbers 100 and 3 mean in this situation.

1. Refer back to your work in the table of the previous task. Use that information and the given coordinate planes to graph the following:

   a. Graph (*t*, *n*) when *t* is 0, 1, 2, 3, and 4.

   b. Graph (*t*, *p*) when *t* is 0, 1, 2, 3, and 4. (If you get stuck, you can create a table.)

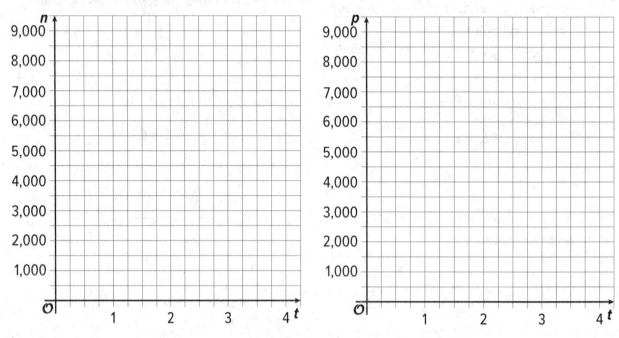

2. On the graph of *n*, where can you see each number that appears in the equation?

3. On the graph of *p*, where can you see each number that appears in the equation?

NAME _____ DATE _____ PERIOD _____

Summary
Representing Exponential Growth

In relationships where the change is exponential, a quantity is repeatedly multiplied by the same amount. The multiplier is called the **growth factor**.

Suppose a population of cells starts at 500 and triples every day. The number of cells each day can be calculated as follows:

Number of Days	Number of Cells
0	500
1	1,500 (or 500 · 3)
2	4,500 (or 500 · 3 · 3, or 500 · 3^2)
3	13,500 (or 500 · 3 · 3 · 3, or 500 · 3^3)
d	500 · 3^d

We can see that the number of cells (p) is changing exponentially, and that p can be found by multiplying 500 by 3 as many times as the number of days (d) since the 500 cells were observed. The *growth factor* is 3. To model this situation, we can write this equation: $p = 500 \cdot 3^d$.

The equation can be used to find the population on any day, including day 0, when the population was first measured. On day 0, the population is $500 \cdot 3^0$. Since $3^0 = 1$, this is 500 · 1 or 500.

Here is a graph of the daily cell population. The point (0, 500) on the graph means that on day 0, the population starts at 500.

Each point is 3 times higher on the graph than the previous point. (1, 1500) is 3 times higher than (0, 500), and (2, 4500) is 3 times higher than (1, 1500).

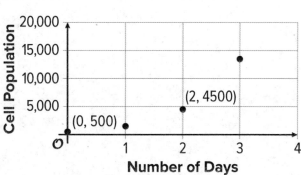

Glossary

growth factor

Practice
Representing Exponential Growth

1. Which expression is equal to $4^0 \cdot 4^2$?

   (A.) 0

   (B.) 1

   (C.) 16

   (D.) 64

2. Select **all** expressions that are equivalent to 3^8. (Lesson 5-1)

   (A.) 8^3

   (B.) $\dfrac{3^{10}}{3^2}$

   (C.) $3 \cdot 8$

   (D.) $(3^4)^2$

   (E.) $(3 \cdot 3)^4$

   (F.) $\dfrac{1}{3^{-8}}$

NAME _____ DATE _____ PERIOD _____

3. A bee population is measured each week and the results are plotted on the graph.

    a. What is the bee population when it is first measured?

    b. Is the bee population growing by the same factor each week? Explain how you know.

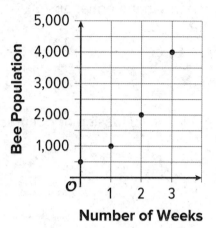

    c. What is an equation that models the bee population, *b*, *w* weeks after it is first measured?

4. A bond is initially bought for \$250. It doubles in value every decade.

    a. Complete the table.

    b. How many decades does it take before the bond is worth more than \$10,000?

    c. Write an equation relating *v*, the value of the bond, to *d*, the number of decades since the bond was bought.

Decades Since Bond is Bought	Dollar Value of Bond
0	
1	
2	
3	
d	

5. A sea turtle population p is modeled by the equation $p = 400 \cdot \left(\frac{5}{4}\right)^y$ where y is the number of years since the population was first measured.

   a. How many turtles are in the population when it is first measured? Where do you see this in the equation?

   b. Is the population increasing or decreasing? How can you tell from the equation?

   c. When will the turtle population reach 700? Explain how you know.

NAME _____ DATE _____ PERIOD _____

6. Bank account A starts with $5,000 and grows by $1,000 each week. Bank account B starts with $1 and doubles each week. **(Lesson 5-1)**

   a. Which account has more money after one week? After two weeks?

   b. Here is a graph showing the two account balances. Which graph corresponds to which situation? Explain how you know.

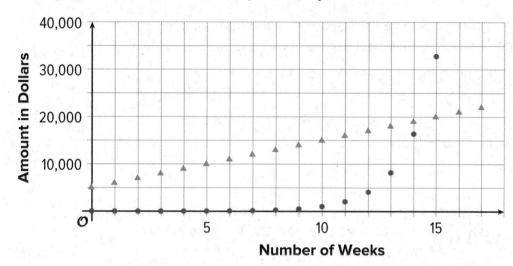

   c. Given a choice, which of the two accounts would you choose? Explain your reasoning.

7. Match each equation in the first list to an equation in the second list that has the same solution. (Lesson 2-9)

A. $y = \frac{2}{5}x + 2$

B. $x = -5 - 2.5y$

C. $y = \frac{10}{5} - 0.4x$

D. $2x = 10 - 5y$

E. $-5y = 2x + 10$

F. $x = 5 - \frac{5}{2}y$

1. $2x + 5y = 10$

2. $-2x - 5y = 10$

3. $-2x + 5y = 10$

8. Function F is defined so that its output $F(t)$ is the number of followers on a social media account t days after set up of the account. (Lesson 4-3)

a. Explain the meaning of $F(30) = 8{,}950$ in this situation.

b. Explain the meaning of $F(0) = 0$.

c. Write a statement about function F that represents the fact that there were 28,800 followers 110 days after the set up of the account.

d. Explain the meaning of t in the equation $F(t) = 100{,}000$.

Lesson 5-4

Understanding Decay

DATE _____ PERIOD _____

NAME _____

Learning Goal Let's look at exponential decay.

 Warm Up

4.1 Notice and Wonder: Two Tables

What do you notice? What do you wonder?

Table A

x	y
0	2
1	$3\frac{1}{2}$
2	5
3	$6\frac{1}{2}$
4	8

Table B

x	y
0	2
1	3
2	$\frac{9}{2}$
3	$\frac{27}{4}$
4	$\frac{81}{8}$

Activity

4.2 What's Left?

1. Here is one way to think about how much Diego has left after spending $\frac{1}{4}$ of $100. Explain each step.

   - Step 1: $100 - \frac{1}{4} \cdot 100$

   - Step 2: $100 \left(1 - \frac{1}{4}\right)$

   - Step 3: $100 \cdot \frac{3}{4}$

   - Step 4: $\frac{3}{4} \cdot 100$

2. A person makes $1,800 per month, but $\frac{1}{3}$ of that amount goes to her rent. What two numbers can you multiply to find out how much she has after paying her rent?

3. Write an expression that only uses multiplication and that is equivalent to x reduced by $\frac{1}{8}$ of x.

Activity

4.3 Value of a Vehicle

Every year after a new car is purchased, it loses $\frac{1}{3}$ of its value. Let's say that a new car costs $18,000.

1. A buyer worries that the car will be worth nothing in three years. Do you agree? Explain your reasoning.

NAME _____ DATE _____ PERIOD _____

2. Write an expression to show how to find the value of the car for each year listed in the table.

Year	Value of Car (Dollars)
0	18,000
1	
2	
3	
6	
t	

3. Write an equation relating the value of the car in dollars, v, to the number of years, t.

4. Use your equation to find v when t is 0. What does this value of v mean in this situation?

5. A different car loses value at a different rate. The value of this different car in dollars, d, after t years can be represented by the equation $d = 10{,}000 \cdot \left(\frac{4}{5}\right)^t$. Explain what the numbers 10,000 and $\frac{4}{5}$ mean in this situation.

Start with an equilateral triangle with area 1 square unit, divide it into 4 congruent pieces as in the figure, and remove the middle one. Then, repeat this process with each of the remaining pieces. Repeat this process over and over for the remaining pieces. The figure shows the first two steps of this construction.

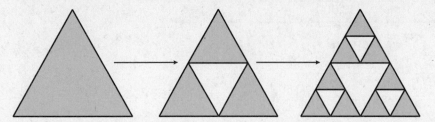

What fraction of the area is removed each time? How much area is removed after the n-th step? Use a calculator to find out how much area *remains* in the triangle after 50 such steps have been taken.

NAME _____ DATE _____ PERIOD _____

Summary
Understanding Decay

Sometimes a quantity grows by the same factor at regular intervals. For example, a population doubles every year. Sometimes a quantity *decreases* by the same factor at regular intervals. For example, a car might lose one third of its value every year.

Let's look at a situation where the quantity decreases by the same factor at regular intervals. Suppose a bacteria population starts at 100,000 and $\frac{1}{4}$ of the population dies each day. The population one day later is $100{,}000 - \frac{1}{4} \cdot 100{,}000$, which can be written as $100{,}000\left(1 - \frac{1}{4}\right)$. The population after one day is $\frac{3}{4}$ of 100,000 or 75,000. The population after two days is $\frac{3}{4} \cdot 75{,}000$. Here are some further values for the bacteria population.

Number of Days	Bacteria Population
0	100,000
1	$75{,}000 \left(\text{or } 100{,}000 \cdot \frac{3}{4}\right)$
2	$56{,}250 \left(\text{or } 100{,}000 \cdot \frac{3}{4} \cdot \frac{3}{4}, \text{ or } 100{,}000 \cdot \left(\frac{3}{4}\right)^2\right)$
3	$\text{about } 42{,}188 \left(\text{or } 100{,}000 \cdot \frac{3}{4} \cdot \frac{3}{4} \cdot \frac{3}{4}, \text{ or } 100{,}000 \cdot \left(\frac{3}{4}\right)^3\right)$

In general, *d* days after the bacteria population was 100,000, the population *p* is given by the equation:

$$p = 100{,}000 \cdot \left(\frac{3}{4}\right)^d,$$

with one factor of $\frac{3}{4}$ for each day.

Situations with quantities that decrease exponentially are described with *exponential decay*. The multiplier $\left(\frac{3}{4} \text{ in this case}\right)$ is still called the *growth factor*, though sometimes people call it the *decay factor* instead.

Glossary

growth factor

Practice

Understanding Decay

1. A new bicycle sells for $300. It is on sale for $\frac{1}{4}$ off the regular price. Select **all** the expressions that represent the sale price of the bicycle in dollars.

(A.) $300 \cdot \frac{1}{4}$

(B.) $300 \cdot \frac{3}{4}$

(C.) $300 \cdot \left(1 - \frac{1}{4}\right)$

(D.) $300 - \frac{1}{4}$

(E.) $300 - \frac{1}{4} \cdot 300$

2. A computer costs $800. It loses $\frac{1}{4}$ of its value every year after it is purchased.

   a. Complete the table to show the value of the computer at the listed times.

Time (years)	Value of Computer (dollars)
0	
1	
2	
3	
t	

   b. Write an equation representing the value, v, of the computer, t years after it is purchased.

   c. Use your equation to find v when t is 5. What does this value of v mean?

NAME _____ DATE _____ PERIOD _____

3. A piece of paper is folded into thirds multiple times. The area, A, of the piece of paper in square inches, after n folds, is $A = 90 \cdot \left(\frac{1}{3}\right)^n$.

   a. What is the value of A when $n = 0$? What does this mean in the situation?

   b. How many folds are needed before the area is less than 1 square inch?

   c. The area of another piece of paper in square inches, after n folds, is given by $B = 100 \cdot \left(\frac{1}{2}\right)^n$. What do the numbers 100 and $\frac{1}{2}$ mean in this situation?

4. At the beginning of April, a colony of ants has a population of 5,000.

   a. The colony decreases by $\frac{1}{5}$ during April. Write an expression for the ant population at the end of April.

   b. During May, the colony decreases again by $\frac{1}{5}$ of its size. Write an expression for the ant population at the end of May.

   c. The colony continues to decrease by $\frac{1}{5}$ of its size each month. Write an expression for the ant population after 6 months.

5. Lin has 13 mystery novels. Each month, she gets 2 more. Select **all** expressions that represent the total number of Lin's mystery novels after 3 months. (Lesson 5-2)

   (A.) $13 + 2 + 2 + 2$

   (B.) $13 \cdot 2 \cdot 2 \cdot 2$

   (C.) $13 \cdot 8$

   (D.) $13 + 6$

   (E.) 19

6. An *odometer* is the part of a car's dashboard that shows the number of miles a car has traveled in its lifetime. Before a road trip, a car odometer reads 15,000 miles. During the trip, the car travels 65 miles per hour. (Lesson 5-2)

Duration of Trip (Hours)	Odometer Reading (Miles)
0	
1	
2	
3	
4	
5	

   a. Complete the table.

   b. What do you notice about the differences of the odometer readings each hour?

   c. If the odometer reads *n* miles at a particular hour, what will it read one hour later?

7. A group of students is collecting 16 oz and 28 oz jars of peanut butter to donate to a food bank. At the end of the collection period, they donated 1,876 oz of peanut butter and a total of 82 jars of peanut butter to the food bank. (Lesson 2-12)

   a. Write a system of equations that represents the constraints in this situation. Be sure to specify the variables that you use.

   b. How many 16 oz jars and how many 28 oz jars of peanut butter were donated to the food bank? Explain or show how you know.

8. A function multiplies its input by $\frac{3}{4}$ then adds 7 to get its output. Use function notation to represent this function. (Lesson 4-4)

9. A function is defined by the equation $f(x) = 2x - 5$. (Lesson 4-4)

   a. What is $f(0)$?   c. What is $f(100)$?

   b. What is $f\left(\frac{1}{2}\right)$?   d. What is x when $f(x) = 9$?

Lesson 5-5

Representing Exponential Decay

NAME _____ DATE _____ PERIOD _____

Learning Goal Let's think about how to show and talk about exponential decay.

Warm Up
5.1 Two Other Tables

Use the patterns you notice to complete the tables. Show your reasoning.

Table A

x	0	1	2	3	4	25
y	2.5	10	17.5	25		

Table B

x	0	1	2	3	4	25
y	2.5	10	40	160		

Activity

5.2 The Algae Bloom

In order to control an algae bloom in a lake, scientists introduce some treatment products.

Once the treatment begins, the area covered by algae A, in square yards, is given by the equation

$A = 240 \cdot \left(\frac{1}{3}\right)^t$. Time, t, is measured in weeks.

1. In the equation, what does the 240 tell us about the algae? What does the $\frac{1}{3}$ tell us?

2. Create a graph to represent $A = 240 \cdot \left(\frac{1}{3}\right)^t$ when t is 0, 1, 2, 3, and 4. Think carefully about how you choose the scale for the axes. If you get stuck, consider creating a table of values.

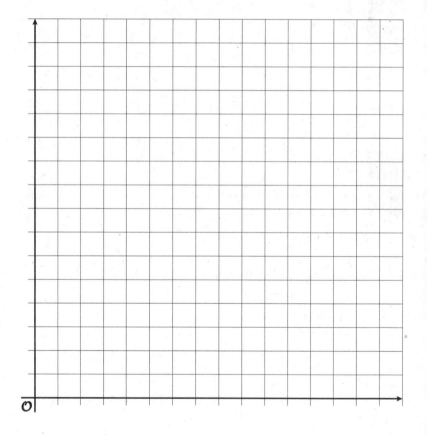

NAME _____ DATE _____ PERIOD _____

3. About how many square yards will the algae cover after 2.5 weeks? Explain your reasoning.

Are you ready for more?

The scientists estimate that to keep the algae bloom from spreading after the treatment concludes, they will need to get the area covered under one square foot. How many weeks should they run the treatment in order to achieve this?

Activity

5.3 Insulin in the Body

A patient who is diabetic receives 100 micrograms of insulin. The graph shows the amount of insulin, in micrograms, remaining in his bloodstream over time, in minutes.

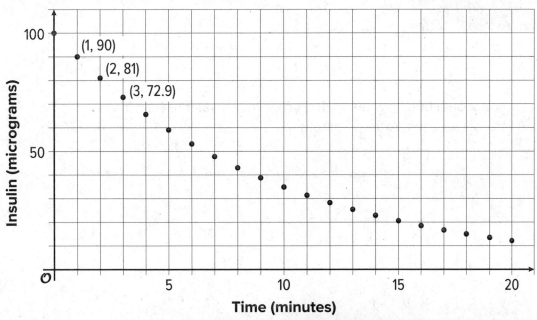

1. Scientists have found that the amount of insulin in a patient's body changes exponentially. How can you check if the graph supports the scientists' claim?

2. How much insulin broke down in the first minute? What fraction of the original insulin is that?

3. How much insulin broke down in the second minute? What fraction is that of the amount one minute earlier?

4. What fraction of insulin remains in the bloodstream for each minute that passes? Explain your reasoning.

5. Complete the table to show the predicted amount of insulin 4 and 5 minutes after injection.

Time after Injection (Minutes)	0	1	2	3	4	5
Insulin in the Bloodstream (Micrograms)	100	90	81	72.9		

6. Describe how you would find how many micrograms of insulin remain in his bloodstream after 10 minutes. After m minutes?

NAME _____ DATE _____ PERIOD _____

Summary

Representing Exponential Decay

Here is a graph showing the amount of caffeine in a person's body, measured in milligrams, over a period of time, measured in hours. We are told that the amount of caffeine in the person's body changes exponentially.

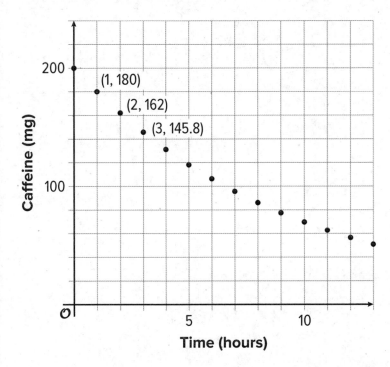

The graph includes the point (0, 200). This means that there were 200 milligrams of caffeine in the person's body when it was initially measured. The point (1, 180) tells us there were 180 milligrams of caffeine 1 hour later. Between 6 and 7 hours after the initial measurement, the amount of caffeine in the body fell below 100 milligrams.

We can use the graph to find out what fraction of caffeine remains in the body each hour. Notice that $\frac{180}{200} = \frac{9}{10}$ and $\frac{162}{180} = \frac{9}{10}$. As each hour passes, the amount of caffeine that stays in the body is multiplied by a factor of $\frac{9}{10}$.

If y is the amount of caffeine, in milligrams, and t is time, in hours, then this situation is modeled by the equation:

$$y = 200 \cdot \left(\frac{9}{10}\right)^t$$

Practice

Representing Exponential Decay

1. A population p of migrating butterflies satisfies the equation
    $p = 100{,}000 \cdot \left(\frac{4}{5}\right)^{w}$ where w is the number of weeks since they
    began their migration.

    a. Complete the table with the population after different numbers of weeks.

w	0	1	2	3	4
p					

    b. Graph the butterfly population.

    Think carefully about how to choose
    a scale for the axes.

    c. What is the vertical intercept of the
    graph? What does it tell you about
    the butterfly population?

    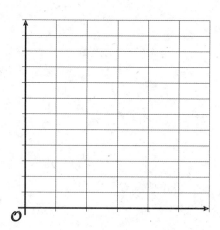

    d. About when does the butterfly population
    reach 50,000?

2. The graph shows the amount of a chemical in a water sample. It is
    decreasing exponentially.

    Find the coordinates of the points labeled A, B,
    and C. Explain your reasoning.

    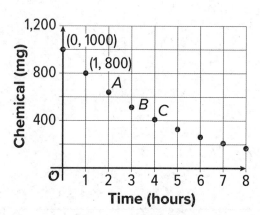

NAME _____ DATE _____ PERIOD _____

3. The graph shows the amount of a chemical in a water sample at different times after it was first measured.

   Select **all** statements that are true.

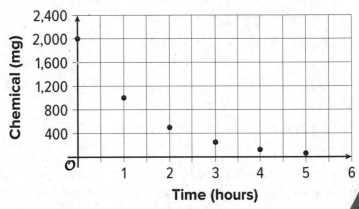

A. The amount of the chemical in the water sample is decreasing exponentially.

B. The amount of the chemical in the water sample is not decreasing exponentially.

C. It is not possible to tell for certain whether or not the amount of the chemical is decreasing exponentially.

D. When it was first measured, there were 2,000 mg of the chemical in the water sample.

E. After 4 hours, there were 100 mg of the chemical in the water.

4. The graph shows the amount of a chemical in a patient's body at different times measured in hours since the levels were first checked.

   Could the amount of this chemical in the patient be decaying exponentially? Explain how you know.

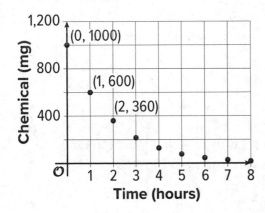

5. The height of a plant in mm is 7. It doubles each week. Select **all** expressions that represent the height of the plant, in mm, after 4 weeks. (Lesson 5-2)

(A.) $7 + 4 \cdot 2$

(D.) $7 \cdot 2 \cdot 2 \cdot 2 \cdot 2$

(B.) $7 \cdot 2^4$

(E.) $7 \cdot 2 \cdot 4$

(C.) $2 + 7^4$

6. The number of people who have read a new book is 300 at the beginning of January. The number of people who have read the book doubles each month. (Lesson 5-2)

   a. Use this information to complete the table.

Number of Months Since January	Number of People Who Have Read the Book
0	
1	
2	
3	
4	

   b. What do you notice about the difference in the number of people who have read the book from month to month?

   c. What do you notice about the factor by which the number of people changes each month?

   d. At the beginning of the month, n people have read the book. How many people will have read the book at the beginning of the next month?

7. Solve each system of equations. (Lesson 2-13)

   a. $\begin{cases} x + y = 2 \\ -3x - y = 5 \end{cases}$

   b. $\begin{cases} \frac{1}{2}x + 2y = -13 \\ x - 4y = 8 \end{cases}$

Lesson 5-6

Analyzing Graphs

NAME _____ DATE _____ PERIOD _____

Learning Goal Let's compare situations where quantities change exponentially.

Warm Up
6.1 Fractions and Decimals

Fraction	$\frac{1}{2}$	$\frac{1}{4}$	$\frac{1}{8}$	$\frac{1}{16}$	$\frac{1}{32}$
Decimal	0.5	0.25	0.125		

In the table, find as many patterns as you can. Use one or more patterns to help you complete the table. Be prepared to explain your reasoning.

Activity

6.2 Falling and Falling

The value of some cell phones changes exponentially after initial release. Here are graphs showing the depreciation of two phones 1, 2, and 3 years after they were released.

Phone A

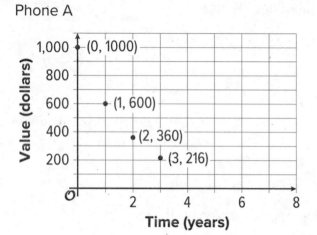

Phone B

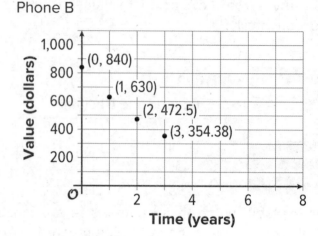

1. Which phone is more expensive to buy when it is first released?

2. How does the value of each phone change with every passing year?

3. Which one is falling in value more quickly? Explain or show how you know.

NAME _____ DATE _____ PERIOD _____

4. If the phones continue to depreciate by the same factor each year, what will the value of each phone be 4 years after its initial release?

5. For each cell phone, write an equation that relates the value of the phone in dollars to the years since release, t. Use v for the value of Phone A and w for the value of Phone B.

Are you ready for more?

When given data, it is not always clear how to best model it. In this case we were told the value of the cell phones was changing exponentially. Suppose, however, we were instead just given the initial values of the cell phones when released and the values after each of the first three years.

1. Use technology to compute the best fit line for each cell phone. Round any numbers to the nearest dollar.

2. Explain why, in this situation, an exponential model might be more appropriate than the linear model you just created.

 Activity

6.3 Card Sort: Matching Descriptions to Graphs

Your teacher will give you a set of cards containing descriptions of situations and graphs. Match each situation with a graph that represents it. Record your matches and be prepared to explain your reasoning.

Summary
Analyzing Graphs

Graphs are useful for comparing relationships. Here are two graphs representing the amount of caffeine in Person A and Person B, in milligrams, at different times, measured hourly, after an initial measurement.

A

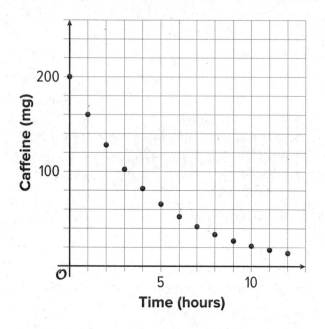

B

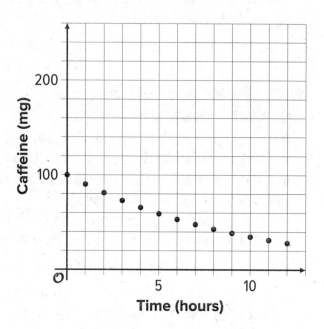

The graphs reveal interesting information about the caffeine in each person over time:

- At the initial measurement, Person A has more caffeine (200 milligrams) than Person B (100 milligrams).

- The caffeine in Person A's body decreases faster. It went from 200 to 160 milligrams in an hour. Because 160 is $\frac{8}{10}$ or $\frac{4}{5}$ of 200, the growth factor is $\frac{4}{5}$.

- The caffeine in Person B's body went from 100 to about 90 milligrams, so that growth factor is about $\frac{9}{10}$. This means that after each hour, a larger fraction of caffeine stays in Person B than in Person A.

- Even though Person A started out with twice as much caffeine, because of the growth factor, Person A had less caffeine than Person B after 6 hours.

NAME _____ DATE _____ PERIOD _____

Practice
Analyzing Graphs

1. The two graphs show models characterized by exponential decay representing the area covered by two different algae blooms, in square yards, *w* weeks after different chemicals were applied.

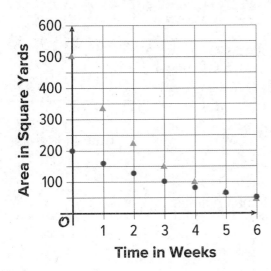

a. Which algae bloom covered a larger area when the chemicals were applied? Explain how you know.

b. Which algae population is decreasing more rapidly? Explain how you know.

2. A medicine is applied to a burn on a patient's arm. The area of the burn in square centimeters decreases exponentially and is shown in the graph.

a. What fraction of the burn area remains each week?

b. Write an equation representing the area of the burn, a, after t weeks.

c. What is the area of the burn after 7 weeks? Round to three decimal places.

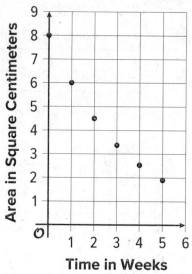

3. Respond to each question.

a. The area of a sheet of paper is 100 square inches. Write an equation that gives the area, A, of the sheet of paper, in square inches, after being folded in half n times.

b. The area of another sheet of paper is 200 square inches. Write an equation that gives the area, B, of this sheet of paper, in square inches, after being folded into thirds n times.

c. Are the areas of the two sheets of paper ever the same after each being folded n times? Explain how you know.

NAME _____ DATE _____ PERIOD _____

4. The graphs show the amounts of medicine in two patients after receiving injections. The circles show the medicine in patient A and the triangles show that in patient B.

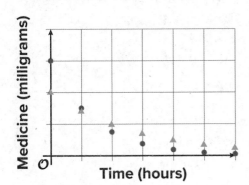

One equation that gives the amount of medicine in milligrams, m, in patient A, h hours after an injection, is $m = 300 \left(\frac{1}{2}\right)^h$.

What could be an equation for the amount of medicine in patient B?

(A.) $m = 500 \left(\frac{3}{10}\right)^h$

(B.) $m = 500 \left(\frac{7}{10}\right)^h$

(C.) $m = 200 \left(\frac{3}{10}\right)^h$

(D.) $m = 200 \left(\frac{7}{10}\right)^h$

5. Select **all** expressions that are equivalent to 3^8. (Lesson 5-3)

   (A.) $3^2 \cdot 3^4$

   (B.) $3^2 \cdot 3^6$

   (C.) $\dfrac{3^{16}}{3^2}$

   (D.) $\dfrac{3^{12}}{3^4}$

   (E.) $(3^4)^2$

   (F.) $(3^1)^7$

6. *Technology required.* Use a graphing calculator to determine the equation of the line of best fit. Round numbers to 2 decimal places. (Lesson 3-5)

x	10	12	15	16	18	20	24
y	27	22	21	19	15	14	10

Lesson 5-7

Using Negative Exponents

NAME _____ DATE _____ PERIOD _____

Learning Goal Let's look more closely at exponential graphs and equations.

 Warm Up
7.1 Exponent Rules

How would you rewrite each of the following as an equivalent expression with a single exponent?

- $2^4 \cdot 2^0$

- $2^4 \cdot 2^{-1}$

- $2^4 \cdot 2^{-3}$

- $2^4 \cdot 2^{-4}$

Activity

7.2 Coral in the Sea

A marine biologist estimates that a structure of coral has a volume of 1,200 cubic centimeters and that its volume doubles each year.

1. Write an equation of the form $y = a \cdot b^t$ representing the relationship, where t is time in years since the coral was measured and y is volume of coral in cubic centimeters. (You need to figure out what numbers a and b are in this situation.)

2. Find the volume of the coral when t is 5, 1, 0, -1, and -2.

3. What does it mean, in this situation, when t is -2?

4. In a certain year, the volume of the coral is 37.5 cubic centimeters. Which year is this? Explain your reasoning.

NAME _____ DATE _____ PERIOD _____

Activity
7.3 Windows of Graphs

The volume, *y*, of coral in cubic centimeters is modeled by the equation
$y = 1{,}200 \cdot 2^x$ where *x* is the number of years since the coral was measured.
Three students used graphing technology to graph the equation that
represents the volume of coral as a function of time.

A

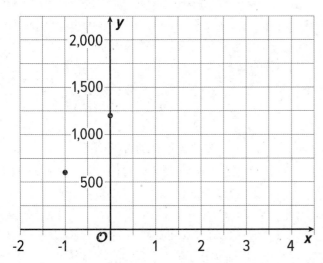

B

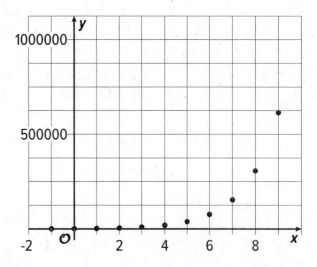

C

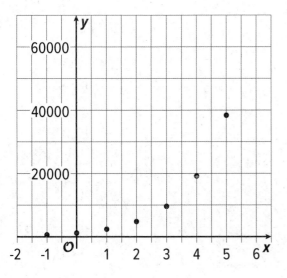

For each graph:

1. Describe how well each graphing window does, or does not, show the behavior of the function.

2. For each graphing window you think does not show the behavior of the function well, describe how you would change it.

3. Make the change(s) you suggested, and sketch the revised graph using graphing technology.

NAME _____ DATE _____ PERIOD _____

Activity
7.4 Measuring Meds

A person took some medicine but does not remember how much. Concerned that she took too much, she has a blood test every hour for several hours.

1. Respond to each question.

   a. Time t is measured in hours since the first blood test and the amount of medicine in her body, m, is measured in milligrams. What is the growth factor? That is, what is b in an equation of the form $m = a \cdot b^t$? What is a?

   b. Find the amounts of medicine in the patient's body when t is -1 and -3. Record them in the table.

t, Time (hours)	m, Medicine (mg)
0	100
1	50
2	25

2. What do $t = 0$ and $t = -3$ mean in this context?

3. The medicine was taken when t is -5. Assuming the person did not have any of the medication in her body beforehand, how much medicine did the patient take?

4. Plot the points whose coordinates are shown in the table. Make sure to draw and label tick marks on the axes.

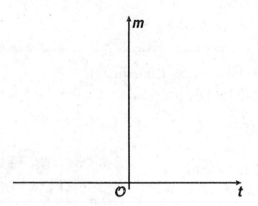

5. Based on your graph, when do you think the patient will have:

   a. 500 mg of medicine remaining in the body

   b. no medicine remaining in the body

Are you ready for more?

Without evaluating them, describe each of the following quantities as close to 0, close to 1, or much larger than 1.

$$\frac{1}{1-2^{-10}} \qquad \frac{2^{10}}{2^{10}+1} \qquad \frac{2^{-10}}{2^{10}+1} \qquad \frac{1-2^{-10}}{2^{10}} \qquad \frac{1+2^{10}}{2^{-10}}$$

NAME _____ DATE _____ PERIOD _____

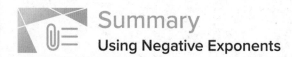

Summary
Using Negative Exponents

Equations are useful not only for representing relationships that change exponentially, but also for answering questions about these situations.

Suppose a bacteria population of 1,000,000 has been increasing by a factor of 2 every hour. What was the size of the population 5 hours ago? How many hours ago was the population less than 1,000?

We could go backwards and calculate the population of bacteria 1 hour ago, 2 hours ago, and so on. For example, if the population doubled each hour and was 1,000,000 when first observed, an hour before then it must have been 500,000, and two hours before then it must have been 250,000, and so on.

Another way to reason through these questions is by representing the situation with an equation. If t measures time in hours since the population was 1,000,000, then the bacteria population can be described by the equation:

$$p = 1,000,000 \cdot 2^t$$

The population is 1,000,000 when t is 0, so 5 hours earlier, t would be -5 and here is a way to calculate the population:

$$1,000,000 \cdot 2^{-5} = 1,000,000 \cdot \frac{1}{2^5}$$

$$= 1,000,000 \cdot \frac{1}{32}$$

$$= 31,250$$

Likewise, substituting -10 for t gives us $1,000,000 \cdot 2^{-10}$ or $1,000,000 \cdot \frac{1}{2^{10}}$, which is a little less than 1,000. This means that 10 hours before the initial measurement the bacteria population was less than 1,000.

Practice

Using Negative Exponents

1. A forest fire has been burning for several days. The burned area, in acres, is given by the equation $y = (4{,}800) \cdot 2^d$, where d is the number of days since the area of the fire was first measured.

    a. Complete the table.

    b. Look at the value of $y = 4{,}800 \cdot 2^d$ when $d = -1$. What does it tell you about the area burned in the fire? What about when $d = -3$?

d, Days Since First Measurement	y, Acres Burned Since Fire Started
0	
-1	
-2	
-3	
-5	

    c. How much area had the fire burned a week before it measured 4,800 acres? Explain your reasoning.

2. The value of a home in 2015 was \$400,000. Its value has been doubling each decade.

    a. If v is the value of the home, in dollars, write an equation for v in terms of d, the number of decades since 2015.

    b. What is v when $d = -1$? What does this value mean?

    c. What is v when $d = -3$? What does this value mean?

3. A fish population, p, can be represented by the equation $p = 800 \cdot \left(\frac{1}{2}\right)^t$ where t is time in years since the beginning of 2015.

    What was the fish population at the beginning of 2012?

    (A.) 100

    (C.) 2,400

    (B.) 800

    (D.) 6,400

NAME _____ DATE _____ PERIOD _____

4. The area, A, of a forest, in acres, is modeled by the equation
$A = 5{,}000 \cdot \left(\dfrac{5}{4}\right)^{d}$ where d is the number of decades since the
beginning of the year 1950.

   a. Is the area of the forest increasing or decreasing with time? Explain how you know.

   b. What was the area of the forest in 1950?

   c. What was the area of the forest in 1940?

   d. Was the area of the forest less than 1,000 acres in 1900? Explain how you know.

5. A population of mosquitos p is modeled by the equation $p = 1{,}000 \cdot 2^{w}$ where w is the number of weeks after the population was first measured. (Lesson 5-3)

   a. Find and plot the mosquito population for $w = 0, 1, 2, 3, 4$.

   b. Where on the graph do you see the 1,000 from the equation for p?

   c. Where on the graph can you see the 2 from the equation?

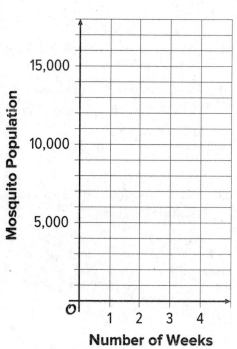

6. The number of copies of a book sold the year it was released was 600,000. Each year after that, the number of copies sold decreased by $\frac{1}{2}$. (Lesson 5-4)

a. Complete the table showing the number of copies of the book sold each year.

b. Write an equation representing the number of copies, c, sold y years after the book was released.

c. Use your equation to find c when $y = 6$. What does this mean in terms of the book?

Years Since Published	Number of Copies Sold
0	
1	
2	
3	
y	

7. The graph shows a population of butterflies, t weeks since their migration began. (Lesson 5-5)

a. How many butterflies were in the population when they started the migration? Explain how you know.

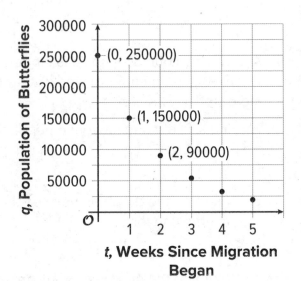

b. How many butterflies were in the population after 1 week? What about after 2 weeks?

c. Write an equation for the population, q, after t weeks.

Lesson 5-8

Exponential Situations as Functions

NAME _____ DATE _____ PERIOD _____

Learning Goal Let's explore exponential functions.

Warm Up
8.1 Rainfall in Las Vegas

Here is a graph of the accumulated rainfall in Las Vegas, Nevada, in the first 60 days of 2017.

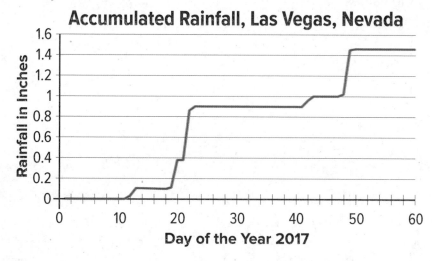

Use the graph to support your answers to the following questions.

1. Is the accumulated amount of rainfall a function of time?

2. Is time a function of accumulated rainfall?

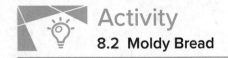

Activity

8.2 Moldy Bread

Clare noticed mold on the last slice of bread in a plastic bag. The area covered by the mold was about 1 square millimeter. She left the bread alone to see how the mold would grow. The next day, the area covered by the mold had doubled, and it doubled again the day after that.

1. If the doubling pattern continues, how many square millimeters will the mold cover 4 days after she noticed the mold? Show your reasoning.

2. Represent the relationship between the area A, in square millimeters, covered by the mold and the number of days d since the mold was spotted using:

   a. A table of values, showing the values from the day the mold was spotted through 5 days later.

   b. An equation

   c. A graph

NAME _____ DATE _____ PERIOD _____

3. Discuss with your partner: Is the relationship between the area covered by mold and the number of days a function? If so, write _____ is a function of _____. If not, explain why it is not.

Are you ready for more?

What do you think an appropriate domain for the mold area function A is? Explain your reasoning.

 Activity

8.3 Functionally Speaking

Here are some situations we have seen previously. For each situation:

- Write a sentence of the form "_____ is a function of _____."
- Indicate which is the independent and which is the dependent variable.
- Write an equation that represents the situation using function notation.

1. In a biology lab, a population of 50 bacteria reproduce by splitting. Every hour, on the hour, each bacterium splits into two bacteria.

2. Every year after a new car is purchased, it loses $\frac{1}{3}$ of its value. Let's say that the new car costs $18,000.

3. In order to control an algae bloom in a lake, scientists introduce some treatment products. The day they begin treatment, the area covered by algae is 240 square yards. Each day since the treatment began, a third of the previous day's area (in square yards) remains covered by algae. Time t is measured in days.

Activity

8.4 Deciding on Graphing Window

The equation $m = 20 \cdot (0.8)^h$ models the amount of medicine m (in milligrams) in a patient's body as a function of hours, h, after injection.

1. Without using a graphing tool, decide if the following horizontal and vertical boundaries are suitable for graphing this function. Explain your reasoning.

$$-10 < h < 100$$

$$-100 < m < 1,000$$

NAME _____ DATE _____ PERIOD _____

2. Verify your answer by graphing the equation using graphing technology, and using the given graphing window. What do you see? Sketch or describe the graph.

3. If your graph in the previous question is unhelpful, modify the window settings so that the graph is more useful. Record the window settings here. Convince a partner why the horizontal and vertical boundaries that you set are better.

Summary

Exponential Situations as Functions

The situations we have looked at that are characterized by exponential change can be seen as *functions*. In each situation, there is a quantity—an independent variable—that determines another quantity—a dependent variable. They are functions because any value of the independent variable corresponds to one and only one value of the dependent variable. Functions that describe *exponential change* are called **exponential functions**.

For example, suppose t represents time in hours and p is a bacteria population t hours after the bacteria population was measured. For each time t, there is only one value for the corresponding number of bacteria, so we can say that p is a function of t and we can write this as $p = f(t)$.

If there were 100,000 bacteria at the time it was initially measured and the population decreases so that $\frac{1}{5}$ of it remains after each passing hour, we can use function notation to model the bacteria population:

$$f(t) = 100,000 \cdot \left(\frac{1}{5}\right)^t$$

Notice the expression in the form of $a \cdot b^t$ (on the right side of the equation) is the same as in previous equations we wrote to represent situations characterized by exponential change.

Glossary

exponential function

NAME _____ DATE _____ PERIOD _____

Practice

Exponential Situations as Functions

1. For an experiment, a scientist designs a can, 20 cm in height, that holds water. A tube is installed at the bottom of the can allowing water to drain out.

   At the beginning of the experiment, the can is full. Every minute after the start of the experiment $\frac{2}{3}$ of the water is drained.

   a. Explain why the height of the water in the can is a function of time.

   b. The height, h, in cm, is a function f of time t in minutes since the beginning of the experiment, $h = f(t)$. Find an expression for $f(t)$.

   c. Find and record the values for f when t is 0, 1, 2, and 3.

   d. Find $f(4)$. What does $f(4)$ represent?

   e. Sketch a graph of f by hand or use graphing technology.

   f. What happens to the level of water in the can as time continues to elapse? How do you see this in the graph?

2. A scientist measures the height, h, of a tree each month, and m is the number of months since the scientist first measured the height of the tree.

    a. Is the height, h, a function of the month, m? Explain how you know.

    b. Is the month, m, a function of the height, h? Explain how you know.

3. A bacteria population is 10,000. It triples each day.

    a. Explain why the bacteria population, b, is a function of the number of days, d, since it was measured to be 10,000.

    b. Which variable is the independent variable in this situation?

    c. Write an equation relating b and d.

4. Respond to each question.

    a. Is the position, p, of the minute hand on a clock a function of the time, t?

    b. Is the time, t, a function of the position of the minute hand on a clock?

NAME _____ DATE _____ PERIOD _____

5. The area covered by a city is 20 square miles. The area grows by a factor of 1.1 each year since it was 20 square miles.

   a. Explain why the area, a, covered by the city, in square miles, is a function of t, the number of years since its area was 20 square miles.

   b. Write an equation for a in terms of t.

6. The graph shows an exponential relationship between x and y. (Lesson 5-7)

   a. Write an equation representing this relationship.

   b. What is the value of y when $x = -1$? Label this point on the graph.

   c. What is the value of y when $x = -2$? Label this point on the graph.

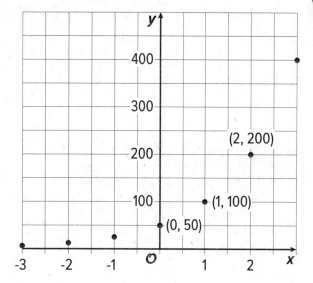

7. Here is an inequality: $3x + 1 > 34 - 4x$. (Lesson 2-19)

   Graph the solution set to the inequality on the number line.

   -10 -9 -8 -7 -6 -5 -4 -3 -2 -1 0 1 2 3 4 5 6 7 8 9 10

8. Here are the equations that define three functions. **(Lesson 4-4)**

$$f(x) = 4x - 5 \qquad g(x) = 4(x - 5) \qquad h(x) = \frac{x}{4} - 5$$

    a. Which function value is the largest: $f(100)$, $g(100)$, or $h(100)$?

    b. Which function value is the largest: $f(-100)$, $g(-100)$, or $h(-100)$?

    c. Which function value is the largest: $f\left(\frac{1}{100}\right)$, $g\left(\frac{1}{100}\right)$, or $h\left(\frac{1}{100}\right)$?

Lesson 5-9

Interpreting Exponential Functions

NAME _____ DATE _____ PERIOD _____

Learning Goal Let's find some meaningful ways to represent exponential functions.

Warm Up
9.1 Equivalent or Not?

Lin and Diego are discussing two expressions: x^2 and 2^x.

- Lin says, "I think the two expressions are equivalent."

- Diego says, "I think the two expressions are only equal for *some* values of x."

Do you agree with either of them? Explain or show your reasoning.

The cost, in dollars, to produce 1 watt of solar energy is a function of the number of years since 1977, t.

From 1977 to 1987, the cost could be modeled by an exponential function f. Here is the graph of the function.

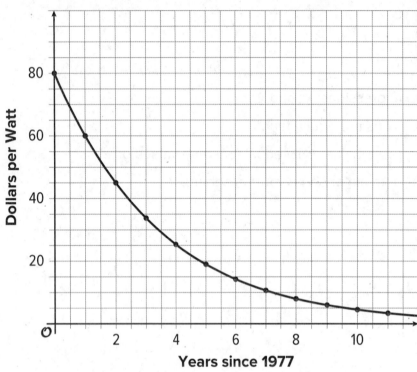

1. What is the statement $f(9) \approx 6$ saying about this situation?

2. What is $f(4)$? What about $f(3.5)$? What do these values represent in this context?

3. When $f(t) = 45$, what is t? What does that value of t represent in this context?

4. By what factor did the cost of solar cells change each year? (If you get stuck, consider creating a table.)

NAME _____ DATE _____ PERIOD _____

Activity

9.3 Paper Folding

1. The thickness t in millimeters of a folded sheet of paper after it is folded n times is given by the equation $t = (0.05) \cdot 2^n$.

   a. What does the number 0.05 represent in the equation?

   b. Use graphing technology to graph the equation $t = (0.05) \cdot 2^n$.

   c. How many folds does it take before the folded sheet of paper is more than 1 mm thick? How many folds before it is more than 1 cm thick? Explain how you know.

2. The area of a sheet of paper is 93.5 square inches.

   a. Find the remaining, visible area of the sheet of paper after it is folded in half once, twice, and three times.

   b. Write an equation expressing the visible area a of the sheet of paper in terms of the number of times it has been folded n.

   c. Use graphing technology to graph the equation.

   d. In this context, can n take negative values? Explain your reasoning.

   e. Can a take negative values? Explain your reasoning.

Are you ready for more?

   1. Using the model in this task, how many folds would be needed to get 1 meter in thickness? 1 kilometer in thickness?

   2. Do some research: what is the current world record for the number of times humans were able to fold a sheet of paper?

NAME _____ DATE _____ PERIOD _____

Activity
9.4 Info Gap: Smartphone Sales

Your teacher will give you either a problem card or a data card. Do not show or read your card to your partner.

If your teacher gives you the data card:

1. Silently read the information on your card.

2. Ask your partner "What specific information do you need?" and wait for your partner to ask for information. Only give information that is on your card. (Do not figure out anything for your partner!)

3. Before telling your partner the information, ask "Why do you need to know (that piece of information)?"

4. Read the problem card, and solve the problem independently.

5. Share the data card, and discuss your reasoning.

If your teacher gives you the problem card:

1. Silently read your card and think about what information you need to answer the question.

2. Ask your partner for the specific information that you need.

3. Explain to your partner how you are using the information to solve the problem.

4. When you have enough information, share the problem card with your partner, and solve the problem independently.

5. Read the data card, and discuss your reasoning.

Pause here so your teacher can review your work. Ask your teacher for a new set of cards and repeat the activity, trading roles with your partner.

Earlier, we used equations to represent situations characterized by exponential change. For example, to describe the amount of caffeine c in a person's body t hours after an initial measurement of 100 mg, we used the equation

$$c = 100 \cdot \left(\frac{9}{10}\right)^t.$$

Notice that the amount of caffeine is a *function* of time, so another way to express this relationship is $c = f(t)$ where f is the function given by

$$f(t) = 100 \cdot \left(\frac{9}{10}\right)^t.$$

We can use this function to analyze the amount of caffeine. For example, when t is 3, the amount of caffeine in the body is $100 \cdot \left(\frac{9}{10}\right)^3$ or $100 \cdot \frac{729}{1,000}$, which is 72.9. The statement $f(3) = 72.9$ means that 72.9 mg of caffeine are present 3 hours after the initial measurement.

We can also graph the function f to better understand what is happening. The point labeled P, for example, has coordinates approximately (10, 35) so it takes about 10 hours after the initial measurement for the caffeine level to decrease to 35 mg.

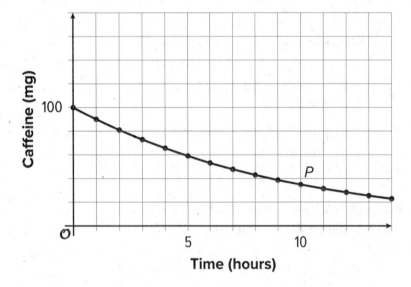

A graph can also help us think about the values in the domain and range of a function. Because the body breaks down caffeine continuously over time, the domain of the function—the time in hours—can include non-whole numbers (for example, we can find the caffeine level when t is 3.5). In this situation, negative values for the domain would represent the time *before* the initial measurement. For example $f(-1)$ would represent the amount of caffeine in the person's body 1 hour before the initial measurement. The range of this function would not include negative values, as a negative amount of caffeine does not make sense in this situation.

NAME _____ DATE _____ PERIOD _____

Practice
Interpreting Exponential Functions

1. The number of people with the flu during an epidemic is a function, f, of the number of days, d, since the epidemic began. The equation $f(d) = 50 \cdot \left(\frac{3}{2}\right)^{d}$ defines f.

   a. How many people had the flu at the beginning of the epidemic? Explain how you know.

   b. How quickly is the flu spreading? Explain how you can tell from the equation.

   c. What does $f(1)$ mean in this situation?

   d. Does $f(3.5)$ make sense in this situation?

2. The function, *f*, gives the dollar value of a bond *t* years after the bond was purchased. The graph of *f* is shown.

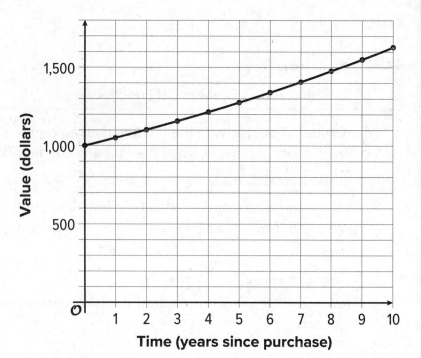

a. What is $f(0)$? What does it mean in this situation?

b. What is $f(4.5)$? What does it mean in this situation?

c. When is $f(t) = 1500$? What does this mean in this situation?

NAME _____ DATE _____ PERIOD _____

3. *Technology required.* A function *f* gives the number of stray cats in a town *t* years since the town started an animal control program. The program includes both sterilizing stray cats and finding homes to adopt them.

   An equation representing *f* is $f(t) = 243 \left(\frac{1}{3}\right)^t$.

   a. What is the value of *f(t)* when *t* is 0? Explain what this value means in this situation.

   b. What is the approximate value of *f(t)* when *t* is $\frac{1}{2}$? Explain what this value means in this situation.

   c. What does the number $\frac{1}{3}$ tell you about the stray cat population?

   d. Use technology to graph *f* for values of *t* between 0 and 4. What graphing window allows you to see values of *f(t)* that correspond to these values of *t*?

4. Function g gives the amount of a chemical in a person's body, in milligrams, t hours since the patient took the drug. The equation $g(t) = 600 \cdot \left(\frac{3}{5}\right)^t$ defines this function.

   a. What does the fraction $\frac{3}{5}$ mean in this situation?

   b. Sketch a graph of g.

   c. What are the domain and range of g? Explain what they mean in this situation.

5. The dollar value of a moped is a function of the number of years, t, since the moped was purchased. The function, f, is defined by the equation $f(t) = 2{,}500 \cdot \left(\frac{1}{2}\right)^t$.

   What is the best choice of domain for the function f?

   (A.) $-10 \leq t \leq 10$

   (B.) $-10 \leq t \leq 0$

   (C.) $0 \leq t \leq 10$

   (D.) $0 \leq t \leq 100$

NAME _____ DATE _____ PERIOD _____

6. A patient receives 1,000 mg of a medicine. Each hour, $\frac{1}{5}$ of the medicine in the patient's body decays. **(Lesson 5-4)**

   a. Complete the table with the amount of medicine in the patient's body.

Hours Since Receiving Medicine	mg of Medicine Left in Body
0	
1	
2	
3	
h	

   b. Write an equation representing the number of mg of the medicine, m, in the patient's body h hours after receiving the medicine.

   c. Use your equation to find m when $h = 10$. What does this mean in terms of the medicine?

7. The trees in a forest are suffering from a disease. The population of trees, p, in thousands, is modeled by the equation $p = 90 \cdot \left(\frac{3}{4}\right)^t$, where t is the number of years since 2000. **(Lesson 5-7)**

   a. What was the tree population in 2001? What about in 1999?

   b. What does the number $\frac{3}{4}$ in the equation for p tell you about the population?

   c. What is the last year when the population was more than 250,000? Explain how you know.

8. All of the students in a classroom list their birthdays. (Lesson 5-8)

    a. Is the birthdate, b, a function of the student, s?

    b. Is the student, s, a function of the birthdate, b?

9. Mai wants to graph the solution to the inequality $5x - 4 > 2x - 19$ on a number line. She solves the equation $5x - 4 = 2x - 19$ for x and gets $x = -5$.

    Which graph shows the solution to the inequality? (Lesson 2-19)

A.
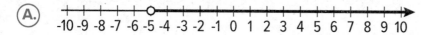

B.
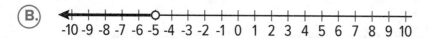

C.
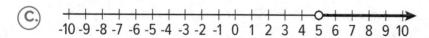

D.
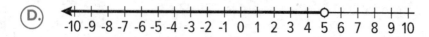

Lesson 5-10

Looking at Rates of Change

NAME _____ DATE _____ PERIOD _____

Learning Goal Let's calculate average rates of change for exponential functions.

 Warm Up

10.1 Falling Prices

Let p be the function that gives the cost $p(t)$, in dollars, of producing 1 watt of solar energy t years after 1977. Here is a table showing the values of p from 1977 to 1987.

t	$p(t)$
0	80
1	60
2	45
3	33.75
4	25.31
5	18.98
6	14.24
7	10.68
8	8.01
9	6.01
10	4.51

Which expression best represents the average rate of change in solar cost between 1977 and 1987?

1. $p(10) - p(0)$

2. $p(10)$

3. $\dfrac{p(10) - p(0)}{10 - 0}$

4. $\dfrac{p(10)}{p(0)}$

Activity

10.2 Coffee Shops

Here is a table and a graph that show the number of coffee shops worldwide that a company had in its first 10 years, between 1987 and 1997. The growth in the number of stores was roughly exponential.

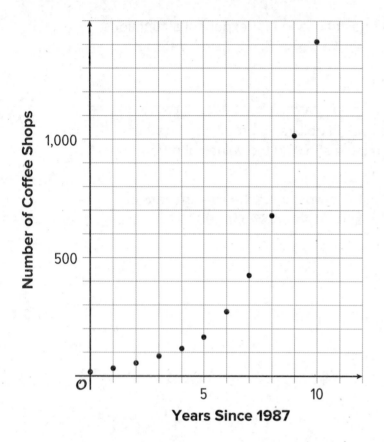

Year	Number of Stores
1987	17
1988	33
1989	55
1990	84
1991	116
1992	165
1993	272
1994	425
1995	677
1996	1,015
1997	1,412

1. Find the average rate of change for each period of time. Show your reasoning.

   a. 1987 and 1990

   b. 1987 and 1993

   c. 1987 and 1997

NAME _____ DATE _____ PERIOD _____

2. Make some observations about the rates of change you calculated. What do these average rates tell us about how the company was growing during this time period?

3. Use the graph to support your answers to these questions. How well do the average rates of change describe the growth of the company in:

   a. the first 3 years?

   b. the first 6 years?

   c. the entire 10 years?

4. Let f be the function so that $f(t)$ represents the number of stores t years since 1987. The value of $f(20)$ is 15,011. Find $\dfrac{f(20) - f(10)}{20 - 10}$ and say what it tells us about the change in the number of stores.

Activity

10.3 Revisiting Cost of Solar Cells

Here is a graph you saw in an earlier lesson. It represents the exponential function p, which models the cost p(t), in dollars, of producing 1 watt of solar energy, from 1977 to 1988 where t is years since 1977.

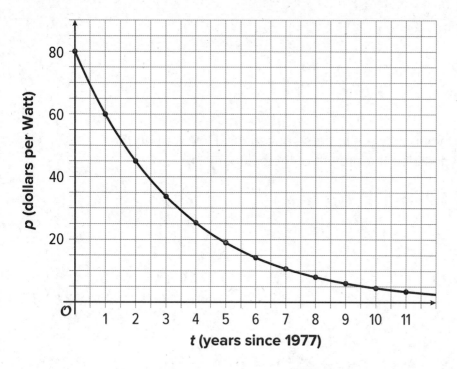

1. Clare said, "In the first five years, between 1977 and 1982, the cost fell by about $12 per year. But in the second five years, between 1983 and 1988, the cost fell only by about $2 a year. Show that Clare is correct.

2. If the trend continues, will the average decrease in price be more or less than $2 per year between 1987 and 1992? Explain your reasoning.

NAME _____ DATE _____ PERIOD _____

Are you ready for more?

Suppose the cost of producing 1 watt of solar energy had instead decreased by $12.20 each year between 1977 and 1982. Compute what the costs would be each year and plot them on the same graph shown in the activity. How do these alternate costs compare to the actual costs shown?

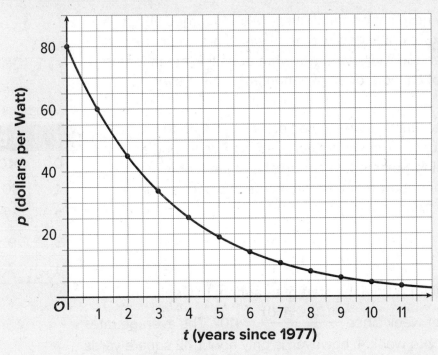

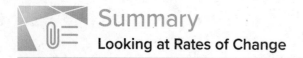

Summary
Looking at Rates of Change

When we calculate the average rate of change for a linear function, no matter what interval we pick, the value of the rate of change is the same. A constant rate of change is an important feature of linear functions! When a linear function is represented by a graph, the slope of the line is the rate of change of the function.

Exponential functions also have important features. We've learned about exponential growth and exponential decay, both of which are characterized by a constant quotient over equal intervals. But what does this mean for the value of the average rate of change for an exponential function over a specific interval?

Let's look at an exponential function we studied earlier. Let A be the function that models the area $A(t)$, in square yards, of algae covering a pond t weeks after beginning treatment to control the algae bloom. Here is a table showing about how many square yards of algae remain during the first 5 weeks of treatment.

t	$A(t)$
0	240
1	80
2	27
3	9
4	3

The average rate of change of A from the start of treatment to week 2 is about -107 square yards per week since $\dfrac{A(2) - A(0)}{2 - 0} \approx -107$. The average rate of change of A from week 2 to week 4, however, is only about -12 square yards per week since $\dfrac{A(4) - A(2)}{4 - 2} \approx -12$.

These calculations show that A is decreasing over both intervals, but the average rate of change is less from weeks 0 to 2 than from weeks 2 to 4, which is due to the effect of the decay factor. If we had looked at an exponential growth function instead, the values for the average rate of change of each interval would be positive with the second interval having a greater value than the first, which is due to the effect of the growth factor.

NAME _____ DATE _____ PERIOD _____

Practice
Looking at Rates of Change

1. A store receives 2,000 decks of popular trading cards. The number of decks of cards is a function, d, of the number of days, t, since the shipment arrived. Here is a table showing some values of d.

   Calculate the average rate of change for the following intervals:

t	$d(t)$
0	2,000
5	1,283
10	823
15	528
20	338

   a. day 0 to day 5

   b. day 15 to day 20

2. A study was conducted to analyze the effects on deer population in a particular area. Let f be an exponential function that gives the population of deer t years after the study began.

   If $f(t) = a \cdot b^t$ and the population is increasing, select **all** statements that must be true.

   (A.) $b > 1$

   (B.) $b < 1$

   (C.) The average rate of change from year 0 to year 5 is less than the average rate of change from year 10 to year 15.

   (D.) The average rate of change from year 0 to year 5 is greater than the average rate of change from year 10 to year 15.

   (E.) $a > 0$

3. Function *f* models the population, in thousands, of a city *t* years after 1930.

The average rate of change of *f* from $t = 0$ to $t = 70$ is approximately 14 thousand people per year.

Is this value a good way to describe the population change of the city over that time period? Explain or show your reasoning.

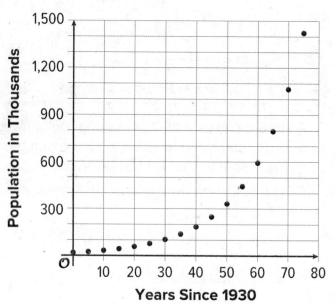

Population in Thousands

Years Since 1930

4. The function, *f*, gives the number of copies a book has sold *w* weeks after it was published. The equation $f(w) = 500 \cdot 2^w$ defines this function.

Select **all** domains for which the average rate of change could be a good measure for the number of books sold.

(A.) $0 \le w \le 2$

(B.) $0 \le w \le 7$

(C.) $5 \le w \le 7$

(D.) $5 \le w \le 10$

(E.) $0 \le w \le 10$

NAME _____ DATE _____ PERIOD _____

5. The graph shows a bacteria population
   decreasing exponentially over time.

   The equation $p = 100{,}000{,}000 \cdot \left(\frac{2}{3}\right)^{h}$ gives the size
   of a second population of bacteria, where h is the
   number of hours since it was measured at 100 million.
   Which bacterial population decays more quickly?
   Explain how you know. (Lesson 5-6)

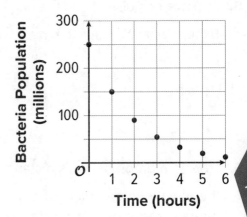

6. *Technology required.* A moth population, p, is modeled by the equation
   $p = 500{,}000 \cdot \left(\frac{1}{2}\right)^{w}$, where w is the number of weeks since the population
   was first measured. (Lesson 5-9)

   a. What was the moth population when it was first measured?

   b. What was the moth population after 1 week? What about 1.5 weeks?

   c. Use technology to graph the population and find out when it falls below 10,000.

7. Give a value for r that indicates that a line of best fit has a positive slope and models the data well. (Lesson 3-7)

8. The size of a district and the number of parks in it have a weak positive relationship.

   Explain what it means to have a weak positive relationship in this context. (Lesson 3-8)

9. Here is a graph of Han's distance from home as he drives.

   Identify the intercepts of the graph and explain what they mean in terms of Han's distance from home. (Lesson 4-6)

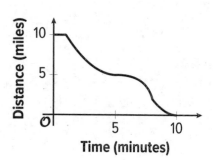

Lesson 5-11

Modeling Exponential Behavior

DATE _____ PERIOD _____

NAME _____

Learning Goal Let's use exponential functions to model real life situations.

Warm Up
11.1 Wondering about Windows

Here is a graph of a function f defined by $f(x) = 400 \cdot (0.2)^x$.

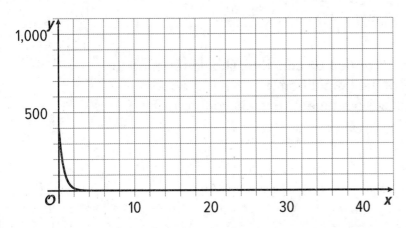

1. Identify the approximate graphing window shown.

2. Suggest a new graphing window that would:

   a. make the graph more informative or meaningful

   b. make the graph less informative or meaningful

Be prepared to explain your reasoning.

Activity

11.2 Beholding Bounces

Here are measurements for the maximum height of a tennis ball after bouncing several times on a concrete surface.

n, Bounce Number	h, Height (Centimeters)
0	150
1	80
2	43
3	20
4	11

1. Which is more appropriate for modeling the maximum height h, in centimeters, of the tennis ball after n bounces: a linear function or an exponential function? Use data from the table to support your answer.

2. Regulations say that a tennis ball, dropped on concrete, should rebound to a height between 53% and 58% of the height from which it is dropped. Does the tennis ball here meet this requirement? Explain your reasoning.

NAME _____ DATE _____ PERIOD _____

3. Write an equation that models the bounce height h after n bounces for this tennis ball.

4. About how many bounces will it take before the rebound height of the tennis ball is less than 1 centimeter? Explain your reasoning.

Activity
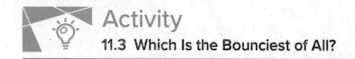

11.3 Which Is the Bounciest of All?

Your teacher will give your group three different kinds of balls.

Your goal is to measure the rebound heights, model the relationship between the number of bounces and the heights, and compare the bounciness of the balls.

1. Complete the table. Make sure to note which ball goes with which column.

n, Number of Bounces	a, Height for Ball 1 (cm)	b, Height for Ball 2 (cm)	c, Height for Ball 3 (cm)
0			
1			
2			
3			
4			

2. Which one appears to be the bounciest? Which one appears to be the least bouncy? Explain your reasoning.

3. For each one, write an equation expressing the bounce height in terms of the bounce number n.

NAME _____ DATE _____ PERIOD _____

4. Explain how the equations could tell us which one is the most bouncy.

5. If the bounciest one were dropped from a height of 300 cm, what equation would model its bounce height *h*?

1. If Ball 1 were dropped from a point that is twice as high, would its bounciness be greater, less, or the same? Explain your reasoning.

2. Ball 4 is half as bouncy as the least bouncy ball. What equation would describe its height *h* in terms of the number of bounces *n*?

3. Ball 5 was dropped from a height of 150 centimeters. It bounced up very slightly once or twice and then began rolling. How would you describe its rebound factor? Explain your reasoning.

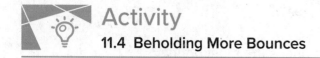
The table shows some heights of a ball after a certain number of bounces.

1. Is this ball more or less bouncy than the tennis ball in the earlier task? Explain or show your reasoning.

Bounce Number	Height in Centimeters
0	
1	
2	73.5
3	51.1
4	36

2. From what height was the ball dropped? Explain or show your reasoning.

3. Write an equation that represents the bounce height of the ball, h, in centimeters after n bounces.

4. Which graph would more appropriately represent the equation for h: Graph A or Graph B? Explain your reasoning.

A

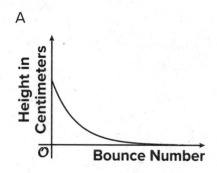

B

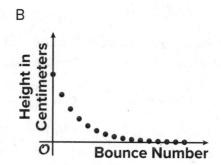

5. Will the n-th bounce of this ball be lower than the n-th bounce of the tennis ball? Explain your reasoning.

NAME _____ DATE _____ PERIOD _____

Summary
Modeling Exponential Behavior

Sometimes data suggest an exponential relationship. For example, this table shows the bounce heights of a certain ball. We can see that the height decreases with each bounce.

To find out what fraction of the height remains after each bounce, we can divide two consecutive values: $\frac{61}{95}$ is about 0.642, $\frac{39}{61}$ is about 0.639, and $\frac{26}{39}$ is about 0.667.

Bounce Number	Bounce Height in Centimeters
1	95
2	61
3	39
4	26

All of these quotients are close to $\frac{2}{3}$. This suggests that there is an exponential relationship between the number of bounces and the height of the bounce, and that the height is decreasing with a factor of about $\frac{2}{3}$ for each successive bounce.

The height h of the ball, in cm, after n bounces can be modeled by the equation:

$$h = 142 \cdot \left(\frac{2}{3}\right)^n$$

Here is a graph of the equation.

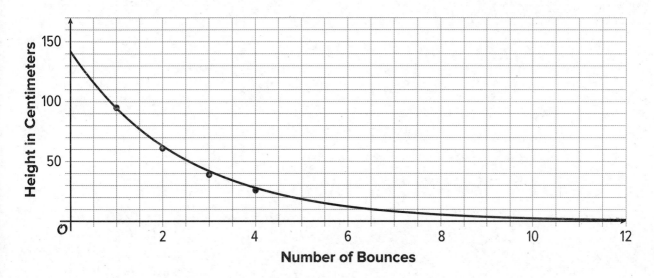

This graph shows both the points from the data and the points generated by the equation, which can give us new insights. For example, the height from which the ball was dropped is not given but can be determined. If $\frac{2}{3}$ of the initial height is about 95 centimeters, then that initial height is about 142.5 centimeters, because $95 \div \frac{2}{3} = 142.5$. For a second example, we can see that it will take 7 bounces before the rebound height is less than 10 centimeters.

Practice

Modeling Exponential Behavior

1. Here is an image showing the highest point of the path of a ball after one bounce.

   Someone is collecting data to model the bounce height of this ball after each bounce. Which measurement for the location of the top of the ball would be the best one to record?

   (A.) 26 cm

   (B.) 26.4 cm

   (C.) 26.43 cm

   (D.) 26.431 cm

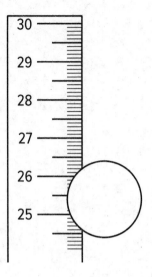

2. Function h describes the height of a ball, in inches, after n bounces and is defined by the equation $h(n) = 120 \cdot \left(\frac{4}{5}\right)^n$

   a. What is $h(3)$? What does it represent in this situation?

   b. Could $h(n)$ be 150? Explain how you know.

   c. Which ball loses its height more quickly, this ball or a tennis ball whose height in inches after n bounces is modeled by the function f where $f(n) = 50 \cdot \left(\frac{5}{9}\right)^n$?

   d. How many bounces would it take before the ball bounces less than 12 inches from the surface?

NAME _____ DATE _____ PERIOD _____

3. After its second bounce, a ball reached a height of 80 cm. The rebound factor for the ball was 0.7. From approximately what height, in cm, was the ball dropped?

(A.) 34

(B.) 49

(C.) 115

(D.) 163

4. Which equation is most appropriate for modeling this data?

(A.) $y = 64 \cdot (1.25)^x$

(B.) $y = 79 \cdot (1.25)^x$

(C.) $y = 79 + 1.25x$

(D.) $y = 64 + 22x$

x	1	2	3	4	5	6
y	79	101	124	158	195	244

5. The table shows the number of employees and number of active customer accounts for some different marketing companies.

Number of Employees	1	2	3	4	10
Number of Customers	4	8	13	17	39

Would a linear or exponential model for the relationship between number of employees and number of customers be more appropriate? Explain how you know.

6. A bank account has a balance of 1,000 dollars. It grows by a factor of 1.04 each year. (Lesson 5-8)

   a. Explain why the balance, in dollars, is a function, f, of the number of years, t, since the account was opened.

   b. Write an equation defining f.

7. The table shows the number of people, n, who went to see a musical on the dth day of April. (Lesson 5-10)

   a. What is the average rate of change for the number of people from day 1 to day 7?

   b. Is the average rate of change a good measure for how the number of people changed throughout the week? Explain your reasoning.

d	n
1	1,534
2	2,324
3	2,418
4	2,281
5	2,350
6	2,394
7	1,720

8. This graph shows the cost (dollars) of mailing a letter from the United States to Canada in 2018 as a function of weight (oz). (Lesson 4-12)

   a. How much does it cost to send a letter that weighs 1.5 oz?

   b. How much does it cost to send a letter that weighs 2 oz?

   c. What is the range of this function?

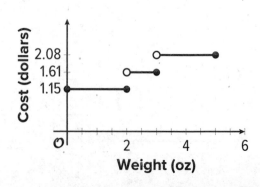

Lesson 5-12

Reasoning about Exponential Graphs (Part 1)

NAME _____ DATE _____ PERIOD _____

Learning Goal Let's study and compare equations and graphs of exponential functions.

 ## Warm Up
12.1 Spending Gift Money

Jada received a gift of $180. In the first week, she spent a third of the gift money. She continues spending a third of what is left each week thereafter. Which equation best represents the amount of gift money g, in dollars, she has after t weeks? Be prepared to explain your reasoning.

(1.) $g = 180 - \frac{1}{3}t$

(2.) $g = 180 \cdot \left(\frac{1}{3}\right)^t$

(3.) $g = \frac{1}{3} \cdot 180^t$

(4.) $g = 180 \cdot \left(\frac{2}{3}\right)^t$

Activity

12.2 Equations and Their Graphs

1. Each of the following functions f, g, h, and j represents the amount of money in a bank account, in dollars, as a function of time x, in years. They are each written in form $m(x) = a \cdot b^x$.

$$f(x) = 50 \cdot 2^x$$
$$g(x) = 50 \cdot 3^x$$
$$h(x) = 50 \cdot \left(\frac{3}{2}\right)^x$$
$$j(x) = 50 \cdot (0.5)^x$$

a. Use graphing technology to graph each function on the same coordinate plane.

b. Explain how changing the value of b changes the graph.

NAME _____ DATE _____ PERIOD _____

2. Here are equations defining functions p, q, and r. They are also written in the form $m(x) = a \cdot b^x$.

$$p(x) = 10 \cdot 4^x$$

$$q(x) = 40 \cdot 4^x$$

$$r(x) = 100 \cdot 4^x$$

a. Use graphing technology to graph each function and check your prediction.

b. Explain how changing the value of a changes the graph.

Are you ready for more?

As before, consider bank accounts whose balances are given by the following functions:

$$f(x) = 10 \cdot 3^x \qquad g(x) = 3^{x+2} \qquad h(x) = \frac{1}{2} \cdot 3^{x+3}$$

Which function would you choose? Does your choice depend on x?

$m(x) = 200 \cdot \left(\frac{1}{4}\right)^x$

$n(x) = 200 \cdot \left(\frac{1}{2}\right)^x$

$p(x) = 200 \cdot \left(\frac{3}{4}\right)^x$

$q(x) = 200 \cdot \left(\frac{7}{8}\right)^x$

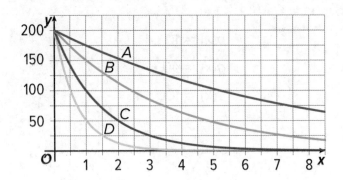

1. Match each equation with a graph. Be prepared to explain your reasoning.

NAME _____ DATE _____ PERIOD _____

2. Functions f and g are defined by these two equations: $f(x) = 1{,}000\left(\frac{1}{10}\right)^x$ and $g(x) = 1{,}000 \cdot \left(\frac{9}{10}\right)^x$.

a. Which function is decaying more quickly? Explain your reasoning.

b. Use graphing technology to verify your response.

An exponential function can give us information about a graph that represents it.

For example, suppose the function q represents a bacteria population t hours after it is first measured and $q(t) = 5{,}000 \cdot (1.5)^t$. The number 5,000 is the bacteria population measured, when t is 0. The number 1.5 indicates that the bacteria population increases by a factor of 1.5 each hour.

A graph can help us see how the starting population (5,000) and growth factor (1.5) influence the population. Suppose functions p and r represent two other bacteria populations and are given by $p(t) = 5{,}000 \cdot 2^t$ and $r(t) = 5{,}000 \cdot (1.2)^t$. Here are the graphs of p, q, and r.

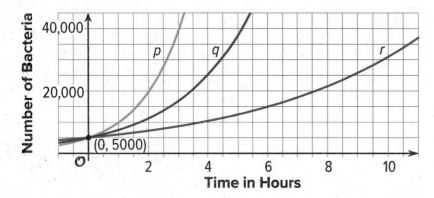

All three graphs start at 5,000 but the graph of r grows more slowly than the graph of q while the graph of p grows more quickly. This makes sense because a population that doubles every hour is growing more quickly than one that increases by a factor of 1.5 each hour, and both grow more quickly than a population that increases by a factor of 1.2 each hour.

NAME _____ DATE _____ PERIOD _____

Practice
Reasoning about Exponential Graphs (Part 1)

1. Here are equations defining three exponential functions f, g, and h.

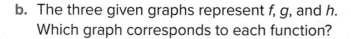

$f(x) = 100 \cdot 3^x$ $\qquad$ $g(x) = 100 \cdot (3.5)^x$ $\qquad$ $h(x) = 100 \cdot 4^x$

a. Which of these functions grows the least quickly? Which one grows the most quickly? Explain how you know.

b. The three given graphs represent f, g, and h. Which graph corresponds to each function?

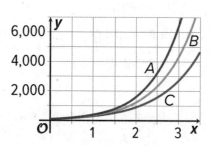

c. Why do all three graphs share the same intersection point with the vertical axis?

2. Here are graphs of three exponential equations.

Match each equation with its graph.

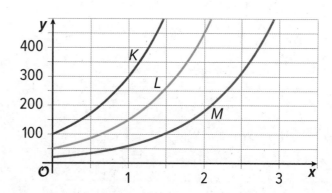

A. $y = 20 \cdot 3^x$ $\qquad\qquad$ 1. K

B. $y = 50 \cdot 3^x$ $\qquad\qquad$ 2. L

C. $y = 100 \cdot 3^x$ $\qquad\qquad$ 3. M

3. The function f is given by $f(x) = 160 \cdot \left(\frac{4}{5}\right)^x$ and the function g is given by $g(x) = 160 \cdot \left(\frac{1}{5}\right)^x$. The graph of f is labeled A and the graph of g is labeled C.

If B is the graph of h and h is defined by $h(x) = a \cdot b^x$, what can you say about a and b? Explain your reasoning.

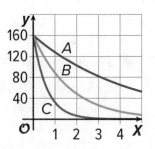

4. Here is a graph of $y = 100 \cdot 2^x$. On the same coordinate plane:

   a. Sketch a graph of $y = 50 \cdot 2^x$ and label it A.

   b. Sketch a graph of $y = 200 \cdot 2^x$ and label it B.

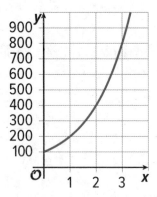

NAME _____ DATE _____ PERIOD _____

5. Choose the inequality whose solution region is represented by this graph. (Lesson 2-21)

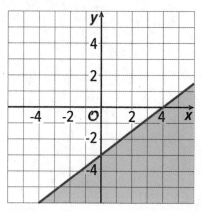

A. $3x - 4y > 12$

B. $3x - 4y \geq 12$

C. $3x - 4y < 12$

D. $3x - 4y \leq 12$

6. *Technology required.* Start with a square with area 1 square unit (not shown). Subdivide it into 9 squares of equal area and remove the middle one to get the first figure shown. (Lesson 5-5)

a. What is the area of the first figure shown?

b. Take the remaining 8 squares, subdivide each into 9 equal squares, and remove the middle one from each. What is the area of the figure now?

c. Continue the process and find the area for stages 3 and 4.

d. Write an equation representing the area *A* at stage *n*.

e. Use technology to graph your equation.

f. Use your graph to find the first stage when the area is first less than $\frac{1}{2}$ square unit.

7. The equation $b = 500 \cdot (1.05)^t$ gives the balance of a bank account t years since the account was opened. The graph shows the annual account balance for 10 years. (Lesson 5-10)

a. What is the average annual rate of change for the bank account?

b. Is the average rate of change a good measure of how the bank account varies? Explain your reasoning.

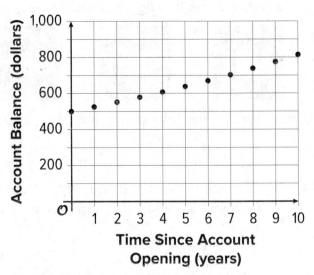

Lesson 5-13

Reasoning about Exponential Graphs (Part 2)

NAME _____ DATE _____ PERIOD _____

Learning Goal Let's investigate what we can learn from graphs that represent exponential functions.

Warm Up
13.1 Which One Doesn't Belong: Four Functions

Which one doesn't belong?

A. $f(n) = 8 \cdot 2^n$

B. $f(n) = 2 \cdot 8^n$

C. $f(n) = 8 + 2n$

D. $f(n) = 8 \cdot \left(\frac{1}{2}\right)^n$

1. Here is a graph representing an exponential function *f*. The function *f* gives the value of a computer, in dollars, as a function of time, *x*, measured in years since the time of purchase.

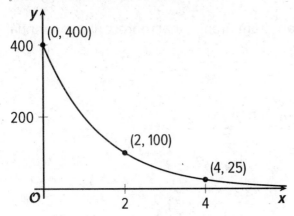

Based on the graph, what can you say about the following?

a. The purchase price of the computer

b. The value of *f* when *x* is 1

c. The meaning of *f*(1)

d. How the value of the computer is changing each year

e. An equation that defines *f*

f. Whether the value of *f* will reach 0 after 10 years

NAME _____ DATE _____ PERIOD _____

2. Here are graphs of two exponential functions. For each, write an equation that defines the function and find the value of the function when x is 5.

a.

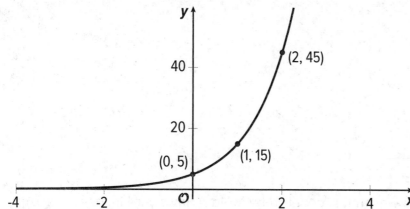

b.

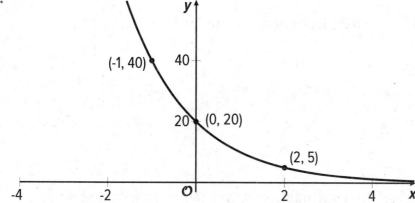

Are you ready for more?

Consider a function f defined by $f(x) = a \cdot b^x$.

- If the graph of f goes through the points (2, 10) and (8, 30), would you expect f(5) to be less than, equal to, or greater than 20?

- If the graph of f goes through the points (2, 30) and (8, 10), would you expect f(5) to be less than, equal to, or greater than 20?

Here are graphs representing two functions, and descriptions of two functions.

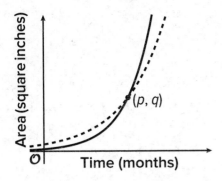

- Function *f*: The area of a wall that is covered by Mold A, in square inches, doubling every month.

- Function *g*: The area of a wall that is covered by Mold B, in square inches, tripling every month.

1. Which graph represents each function? Label the graphs accordingly and explain your reasoning.

2. When the mold was first spotted and measured, was there more of Mold A or Mold B? Explain how you know.

3. What does the point (p, q) tell us in this situation?

NAME _____ DATE _____ PERIOD _____

Summary
Reasoning about Exponential Graphs (Part 2)

If we have enough information about a graph representing an exponential function f, we can write a corresponding equation. Here is a graph of $y = f(x)$.

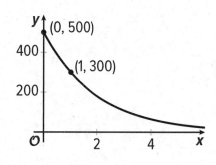

An equation defining an exponential function has the form $f(x) = a \cdot b^x$. The value of a is the starting value or $f(0)$, so it is the y-intercept of the graph. We can see that $f(0)$ is 500 and that the function is decreasing.

The value of b is the growth factor. It is the number by which we multiply the function's output at x to get the output at $x + 1$. To find this growth factor for f, we can calculate $\frac{f(1)}{f(0)}$, which is $\frac{300}{500}$ or $\frac{3}{5}$. So an equation that defines f is:

$$f(x) = 500 \cdot \left(\frac{3}{5}\right)^x$$

We can also use graphs to compare functions. Here are graphs representing two different exponential functions, labeled g and h. Each one represents the area of algae (in square meters) in a pond, x days after certain fish were introduced.

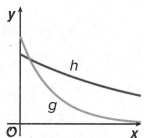

- Pond A had 40 square meters of algae. Its area shrinks to $\frac{8}{10}$ of the area on the previous day.

- Pond B had 50 square meters of algae. Its area shrinks to $\frac{2}{5}$ of the area on the previous day.

Can you tell which graph corresponds to which algae population?

We can see that the y-intercept of g's graph is greater than the y-intercept of h's graph. We can also see that g has a smaller growth factor than h because as x increases by the same amount, g is retaining a smaller fraction of its value compared to h. This suggests that g corresponds to Pond B and h corresponds to Pond A.

1. Here is a graph of p, an insect population, w weeks after it was first measured. The population grows exponentially.

   a. What is the weekly factor of growth for the insect population?

   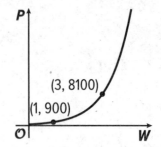

   b. What was the population when it was first measured?

   c. Write an equation relating p and w.

2. Here is a graph of the function f defined by $f(x) = a \cdot b^x$.

   Select **all** possible values of b.

   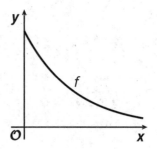

   A. 0

   B. $\frac{1}{10}$

   C. $\frac{1}{2}$

   D. $\frac{9}{10}$

   E. 1

   F. 1.3

   G. $\frac{18}{5}$

NAME _____ DATE _____ PERIOD _____

3. The function f is given by $f(x) = 50 \cdot \left(\frac{1}{2}\right)^x$, and

the function g is given by $g(x) = 50 \cdot \left(\frac{1}{3}\right)^x$.

Here are graphs of f and g.

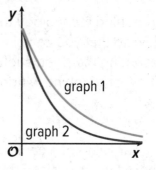

Kiran says that since $3 > 2$, the graph of g lies above the graph of f so graph 1 is the graph of g and graph 2 is the graph of f.

Do you agree? Explain your reasoning.

4. The function f is defined by $f(x) = 50 \cdot 3^x$. The function g is defined by $g(x) = a \cdot b^x$.

Here are graphs of f and g.

a. How does a compare to 50? Explain how you know.

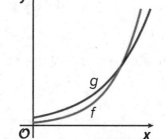

b. How does b compare to 3? Explain how you know.

5. *Technology required.* The equation $y = 600{,}000 \cdot (1.055)^t$ represents the population of a country t decades after the year 2000. (Lesson 5-7)

   Use graphing technology to graph the equation. Then, set the graphing window so that you can simultaneously see points on the graph representing the population predicted by the model in 1980 and in the year 2020. What graphing window did you use?

6. The dollar value of a car is a function, f, of the number of years, t, since the car was purchased. The function is defined by the equation

   $f(t) = 12{,}000 \cdot \left(\dfrac{3}{4}\right)^t$. (Lesson 5-9)

   a. How much was the car worth when it was purchased? Explain how you know.

   b. What is $f(2)$? What does this tell you about the car?

   c. Sketch a graph of the function f.

   d. About when was the car worth $6,000? Explain how you know.

NAME _____ DATE _____ PERIOD _____

7. A ball was dropped from a height of 150 cm. The rebound factor of the ball is 0.8. About how high, in centimeters, did the ball go after the third bounce? (Lesson 5-11)

(A.) 77

(B.) 96

(C.) 234

(D.) 293

8. A triathlon athlete runs at an average rate of 8.2 miles per hour, swims at an average rate of 2.4 miles per hour, and bikes at an average rate of 16.1 miles per hour. At the end of one training session (during which she did not run), she has swum and biked more than 20 miles in total. (Lesson 2-22)

   a. Is it possible that she swam and biked for the following amounts of time in that session? Show your reasoning.

   i. Swam for 0.5 hour and biked 1.25 hours

   ii. Swam for $\frac{1}{3}$ hour and biked for 70 minutes

   b. Write an inequality to represent the relationship between the time she swam and biked, in hours, and the total distance she traveled. Be sure to specify what each variable represents.

c. Use your inequality to graph a solution set that represents all the possible combinations of swimming and running times that meet the distance constraint (regardless of whether the times are realistic).

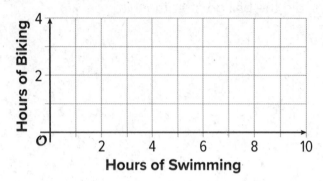

Lesson 5-14

Recalling Percent Change

NAME _____ DATE _____ PERIOD _____

Learning Goal Let's find the result of changing a number by a percentage.

 ## Warm Up
14.1 Wheels

A scooter costs $160.

For each question, show your reasoning.

1. The cost of a pair of roller skates is 20% of the cost of the scooter. How much do the roller skates cost?

2. A bicycle costs 20% more than the scooter. How much does the bicycle cost?

3. A skateboard costs 25% less than the bicycle. How much does the skateboard cost?

1. You need to pay 8% tax on a car that costs $12,000. What will you end up paying in total? Show your reasoning.

2. Burritos are on sale for 30% off. Your favorite burrito normally costs $8.50. How much does it cost now? Show your reasoning.

3. A pair of shoes that originally cost $79 are on sale for 35% off. Does the expression 0.65(79) represent the sale price of the shoes (in dollars)? Explain your reasoning.

Are you ready for more?

Come up with some strategies for mentally adding 15% to the total cost of an item.

NAME _____ DATE _____ PERIOD _____

Activity

14.3 Expressing Percent Increase and Decrease

Complete the table so that each row has a description and two different expressions that answer the question asked in the description. The second expression should use only multiplication. Be prepared to explain how the two expressions are equivalent.

Description and Question	Expression 1	Expression 2 (Using Only Multiplication)
A one-night stay at a hotel in Anaheim, CA costs $160. Hotel room occupancy tax is 15%. What is the total cost of a one-night stay?	$160 + (0.15) \cdot 160$	
Teachers receive 30% educators discount at a museum. An adult ticket costs $24. How much would a teacher pay for admission into the museum?		$(0.7) \cdot 24$
The population of a city was 842,000 ten years ago. The city now has 2% more people than it had then. What is the population of the city now?		
After a major hurricane, 46% of the 90,500 households on an island lost their access to electricity. How many households still have electricity?		
	$754 - (0.21) \cdot 754$	
Two years ago, the number of students in a school was 150. Last year, the student population increased 8%. This year, it increased about 8% again. What is the number of students this year?		

Summary
Recalling Percent Change

We can write different expressions to calculate percent increase and decrease.

Suppose a new phone costs $360 and is on sale at 25% off the regular price. One way to calculate this is to first find 25% of 360, which is 90, and then subtract $90 from $360 to get a sale price of $270. These calculations can be recorded in this way:

$$360 - (0.25) \cdot 360 = 270$$

Another way to represent this calculation is to notice that subtracting 25% of the cost is equivalent to finding 75% of the cost. Using the distributive property, we know that $360 - (0.25) \cdot 360$ can be rewritten as $(1 - 0.25) \cdot 360$, which is equal to $(0.75) \cdot 360$.

NAME _____ DATE _____ PERIOD _____

Practice
Recalling Percent Change

1. For each situation, write an expression answering the question. The expression should only use multiplication.

    a. A person's salary is $2,500 per month. She receives a 10% raise. What is her new salary, in dollars per month?

    b. A test had 40 questions. A student answered 85% of the questions correctly. How many questions did the student answer correctly?

    c. A telephone cost $250. The sales tax is 7.5%. What was the cost of the telephone including sales tax?

2. In June, a family used 3,500 gallons of water. In July, they used 15% more water.

    Select **all** the expressions that represent the number of gallons of water the family used in July.

    (A.) $3,500 + 0.15 \cdot 3,500$

    (B.) $3,500 + 0.15$

    (C.) $3,500 \cdot (1 - 0.15)$

    (D.) $3,500 \cdot (1.15)$

    (E.) $3,500 \cdot (1 + 0.15)$

3. Han's summer job paid him $4,500 last summer. This summer, he will get a 25% pay increase from the company.

   Write two different expressions that could be used to find his new salary, in dollars.

4. Respond to each question.

   a. Military veterans receive a 25% discount on movie tickets that normally cost $16. Explain why 16(0.75) represents the cost of a ticket using the discount.

   b. A new car costs $15,000 and the sales tax is 8%. Explain why 15,000(1.08) represents the cost of the car including tax.

5. The number of grams of a chemical in a pond is a function of the number of days, d, since the chemical was first introduced. The function, f, is defined by $f(d) = 550 \cdot \left(\frac{1}{2}\right)^d$. (Lesson 5-10)

   a. What is the average rate of change between day 0 and day 7?

   b. Is the average rate of change a good measure for how the amount of the chemical in the pond has changed over the week? Explain your reasoning.

NAME _____ DATE _____ PERIOD _____

6. A piece of paper is 0.004 inches thick. **(Lesson 5-8)**

    a. Explain why the thickness in inches, t, is a function of the number of times the paper is folded, n.

    b. Using function notation, represent the relationship between t and n. That is, find a function f so that $t = f(n)$.

7. The function f represents the amount of a medicine, in mg, in a person's body t hours after taking the medicine. Here is a graph of f. **(Lesson 5-13)**

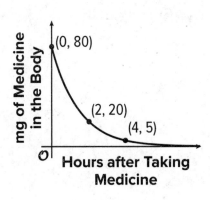

    a. How many mg of the medicine did the person take?

    b. Write an equation that defines f.

    c. After 7 hours, how many mg of medicine remain in the person's body?

8. Match each inequality to the graph of its solution. (Lesson 2-23)

A. $3x + 4y \leq 36$

2.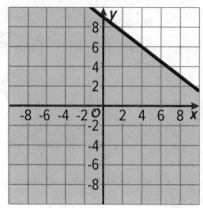

1.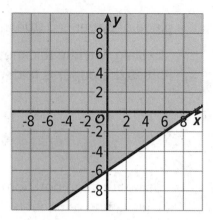

B. $12x + 3y \leq 36$

C. $6x + 4y \geq 36$

3.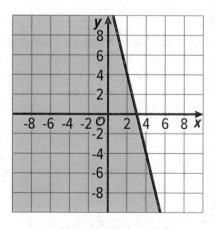

D. $3x - 9y \geq 36$

E. $4x - 6y \leq 36$

4.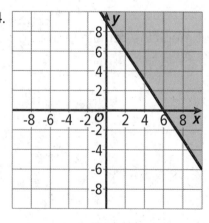

5.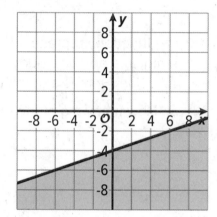

Lesson 5-15

Functions Involving Percent Change

NAME _____ DATE _____ PERIOD _____

Learning Goal Let's investigate what happens when we repeatedly apply a percent increase to a quantity.

Warm Up
15.1 Dandy Discounts

All books at a bookstore are 25% off. Priya bought a book originally priced at $32. The cashier applied the storewide discount and then took another 25% off for a coupon that Priya brought. If there was no sales tax, how much did Priya pay for the book? Show your reasoning.

Activity
15.2 Owing Interests

To get a new computer, a recent college graduate obtains a loan of $450. She agrees to pay 18% annual interest, which will apply to any money she owes. She makes no payments during the first year.

1. How much will she owe at the end of one year? Show your reasoning.

2. Assuming she continues to make no payments to the lender, how much will she owe at the end of two years? Three years?

3. To find the amount owed at the end of the third year, a student started by writing:

$$[\text{Year 3 Amount}] = [\text{Year 2 Amount}] + [\text{Year 2 Amount}] \cdot (0.18)$$
$$\text{and ended with}$$
$$= 450 \cdot (1.18) \cdot (1.18) \cdot (1.18).$$

Does her final expression correctly reflect the amount owed at the end of the third year? Explain or show your reasoning.

4. Write an expression for the amount she owes at the end of x years without payment.

NAME _____ DATE _____ PERIOD _____

Are you ready for more?

Start with a line segment of length 1 unit. Make a new shape by taking the middle third of the line segment and replacing it by two line segments of the same length to reconnect the two pieces. Repeat this process over and over, replacing the middle third of each of the remaining line segments with two segments each of the same length as the segment they replaced, as shown in the figure.

What is the total length of the figure after one iteration of this process (the second shape in the diagram)? After two iterations? After n iterations? Experiment with the value of your expression for large values of n.

Activity

15.3 Comparing Loans

Suppose three people each have taken loans of $1,000 but they each pay different annual interest rates.

1. For each loan, write an expression, using only multiplication, for the amount owed at the end of each year if no payments are made.

Years Without Payment	Loan A 12%	Loan B 24%	Loan C 30.6%
1			
2			
3			
10			
X			

2. Use graphing technology to plot the graphs of the account balances.

3. Based on your graph, about how many years would it take for the original unpaid balance of each loan to double?

NAME _____ DATE _____ PERIOD _____

Activity

15.4 Comparing Average Rates of Change

The functions a, b, and c represent the amount owed (in dollars) for Loans A, B, and C respectively: the input for the functions is t, the number of years without payments.

1. For each loan, find the average rate of change per year between:

   a. the start of the loan and the end of the second year

   b. the end of the tenth year and the end of the twelfth year

2. How do the average rates of change for the three loans compare in each of the two-year intervals?

Summary

Functions Involving Percent Change

When we borrow money from a lender, the lender usually charges *interest*, a percentage of the borrowed amount as payment for allowing us to use the money. The interest is usually calculated at a regular interval of time (monthly, yearly, etc.).

Suppose you received a loan of $500 and the interest rate is 15%, calculated at the end of each year. If you make no other purchases or payments, the amount owed after one year would be $500 + (0.15) \cdot 500$, or $500 \cdot (1 + 0.15)$. If you continue to make no payments or other purchases in the second year, the amount owed would increase by another 15%. The table shows the calculation of the amount owed for the first three years.

Time in Years	Amount Owed in Dollars
1	$500 \cdot (1 + 0.15)$
2	$500 \cdot (1 + 0.15)(1 + 0.15)$, or $500 \cdot (1 + 0.15)^2$
3	$500 \cdot (1 + 0.15)(1 + 0.15)(1 + 0.15)$, or $500 \cdot (1 + 0.15)^3$

The pattern here continues. Each additional year means multiplication by another factor of $(1 + 0.15)$. With no further purchases or payments, after t years the debt in dollars is given by the expression:

$$500 \cdot (1 + 0.15)^t$$

Since exponential functions eventually grow very quickly, leaving a debt unpaid can be very costly.

NAME _____ DATE _____ PERIOD _____

Practice
Functions Involving Percent Change

1. In 2011, the population of deer in a forest was 650.

   a. In 2012, the population increases by 15%. Write an expression, using only multiplication, that represents the deer population in 2012.

   b. In 2013, the population increases again by 15%. Write an expression that represents the deer population in 2013.

   c. If the deer population continues to increase by 15% each year, write an expression that represents the deer population t years after 2011.

2. Mai and Elena are shopping for back-to-school clothes. They found a skirt that originally cost $30 on a 15% off sale rack. Today, the store is offering an additional 15% off. To find the new price of the skirt, in dollars, Mai says they need to calculate $30 \cdot 0.85 \cdot 0.85$. Elena says they can just multiply $30 \cdot 0.70$.

   a. How much will the skirt cost using Mai's method?

   b. How much will the skirt cost using Elena's method?

   c. Explain why the expressions used by Mai and Elena give different prices for the skirt. Which method is correct?

3. *Technology required.* One $1,000 loan charges 5% interest at the end of each year, while a second loan charges 8% interest at the end of each year.

t, Number of Years	b, Loan Balance with 5% Interest	c, Loan Balance with 8% Interest
1		
2		
3		
t		

a. Complete the table with the balances for each loan. Assume that no payments are made and that the interest applies to the entire loan balance, including any previous interest charges.

b. Which loan balance grows more quickly? How will this be visible in the graphs of the two loan balances b and c as functions of the number of years t?

c. Use technology to create graphs representing b and c over time. The graph should show the starting balance of each loan as well as the amount of the loan after 15 years. Write down the graphing window needed to show these points.

4. Lin opened a savings account that pays $5\frac{1}{4}$% interest annually and deposited $5,000.

If she makes no deposits and no withdrawals for 3 years, how much money will be in her account?

5. A person loans his friend $500. They agree to an annual interest rate of 5%.

Write an expression for computing the amount owed on the loan, in dollars, after t years if no payments are made.

NAME _____ DATE _____ PERIOD _____

6. Select **all** situations that are accurately described by the expression $15 \cdot 3^5$.
   (Lesson 5-3)

   (A.) A population of bacteria begins at 15,000. The population triples each hour. How many bacteria are there after 5 hours?

   (B.) A population of bacteria begins at 15,000. The population triples each hour. How many thousand bacteria are there after 5 hours?

   (C.) A population of bacteria begin at 15,000. The population quintuples each hour. How many thousand bacteria are there after 3 hours?

   (D.) A bank account balance is $15. The account balance triples each year. What is the bank account balance, in dollars, after 5 years?

   (E.) A bank account balance is $15,000. It grows by $3,000 each year. What is the bank account balance, in thousands of dollars, after 5 years?

7. Here are graphs of two exponential functions, f and g.

   If $f(x) = 100 \cdot \left(\dfrac{2}{3}\right)^x$ and $g(x) = 100 \cdot b^x$, what could be the value of b? (Lesson 5-13)

   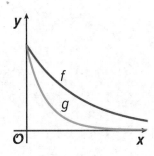

   (A.) $\dfrac{1}{3}$

   (B.) $\dfrac{3}{4}$

   (C.) 1

   (D.) $\dfrac{3}{2}$

8. The real estate tax rate in 2018 in a small rural county increased by $\frac{1}{4}$%. In 2017, a family paid $1,200.

Which expression represents the real estate tax, in dollars, that the family will pay in 2018? **(Lesson 5-14)**

(A.) $1,200 + 1,200 \cdot \left(\frac{1}{4}\right)$

(B.) $1,200 \cdot (1.25)$

(C.) $1,200 \cdot (1.025)$

(D.) $1,200 \cdot (1.0025)$

9. Two inequalities are graphed on the same coordinate plane.

Select **all** of the points that are solutions to the system of the two inequalities. **(Lesson 2-24)**

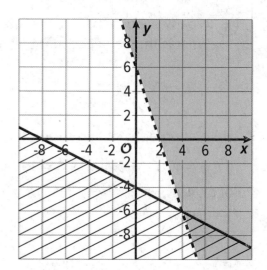

(A.) (4, -6)

(B.) (4, 6)

(C.) (-4, -6)

(D.) (-4, 6)

(E.) (6, -8)

(F.) (7, -9)

(G.) (-8, 6)

Lesson 5-16

Compounding Interest

NAME _____ DATE _____ PERIOD _____

Learning Goal Let's explore different ways of repeatedly applying a percent increase.

Warm Up
16.1 Five Years Later

You owe 12% interest each year on a $500 loan. If you make no payments and take no additional loans, what will the loan balance be after 5 years?

Write an expression to represent the balance and evaluate it to find the answer in dollars.

Activity

16.2 Resizing Images

Andre and Mai need to enlarge two images for a group project. The two images are the same size.

Andre makes a scaled copy of his image, increasing the lengths by 10%. It was still a little too small, so he increases the lengths by 10% again.

Mai says, "If I scale my image and increase the lengths by 20%, our images will be exactly the same size."

Do you agree with Mai? Explain or show your reasoning.

NAME _____ DATE _____ PERIOD _____

Are you ready for more?

Adding 0.01 to a number 10 times is different from multiplying that number by 1.01 ten times, but the values you get from expressions like these are still quite close to one another. For example, $5 + (0.01) \cdot 10 = 5.1$ whereas $5 \cdot (1.01)^{10} \approx 5.523$. The first of these you could do in your head, whereas the second one would probably require a calculator, and so this introduces a neat mental math strategy.

- The quantity $(1.02)^7$ is hard to calculate by hand. Use mental math to compute $1 + (0.02) \cdot 7$ to get a good approximation to it.

- Estimate $(0.99)^9$. Use a calculator to compare your estimate to the actual value.

- Estimate $(1.6)^{11}$. Use a calculator to compare your estimate to the actual value. What is different about this example?

Activity

16.3 Earning Interest

A bank account has a monthly interest rate of 1% and initial balance of $1,000. Any earned interest is added to the account and no other deposits or withdrawals are made.

1. What is the account balance after 6 months, 1 year, 2 years, and 5 years? Show your reasoning.

2. Write an equation expressing the account balance (a) in terms of the number of months (m). Assume that all interest earned continues to be added to the account and no other deposits or withdrawals are made.

3. How much interest will the account earn in 1 year? What percentage of the initial balance is that? Show your reasoning.

4. The term annual return refers to the percent of interest an account holder could expect to receive in one year. Discuss with your partner: If you were the bank, would you advertise the account as having a 12% annual return? Why or why not? Use your work so far to explain your reasoning.

NAME _____ DATE _____ PERIOD _____

Summary
Compounding Interest

Suppose a runner runs 4 miles a day this month. She is increasing her daily running distance by 25% next month, and then by 25% of that the month after. Will she be running 50% more than her current daily distance two months from now?

It is tempting to think that two months from now she will be running 6 miles, since twice of 25% is 50%, and 50% more than her current daily distance is $4 \cdot (1.5)$. But if we calculate the increase one month at a time, we can see that next month she will run $4 \cdot (1.25)$ or 5 miles. The month after that she will run $5 \cdot (1.25)$ or 6.25 miles.

So two months from now her daily distance will actually be:

$$4 \cdot (1.25)^2.$$

Two repeated 25% increases actually lead to an overall increase of 56.25% rather than of 50%, because $1.25^2 = 1.5625$. Applying a percent increase on an amount that has had a prior percent increase is called *compounding*.

Compounding happens when we calculate interest on money in a bank account or on a loan. An account that earns 2% interest every month does not actually earn 24% a year. Let's say a savings account has $300 and no other deposits or withdrawals are made. The account balances after some months are shown in this table.

Number of Months	Account Balance in Dollars
1	$300 \cdot (1.02)$
2	$300 \cdot (1.02)^2$
3	$300 \cdot (1.02)^3$
12	$300 \cdot (1.02)^{12}$

$(1.02)^{12} \approx 1.2682$, so the account will grow by about 26.82% in one year. This rate is called the *effective interest rate*. It reflects how the account balance actually changes after one year.

The 24% is called the *nominal interest rate*. It is the stated or published rate and is usually used to determine the monthly, weekly, or daily rates (if interest were to be calculated at those intervals).

Practice
Compounding Interest

1. Automobiles start losing value, or depreciating, as soon as they leave the car dealership. Five years ago, a family purchased a new car that cost $16,490.

   If the car lost 13% of its value each year, what is the value of the car now?

2. The number of trees in a rainforest decreases each month by 0.5%. The forest currently has 2.5 billion trees.

   Write an expression to represent how many trees will be left in 10 years. Then, evaluate the expression.

3. From 2005 to 2015, a population of P lions is modeled by the equation $P = 1,500 \cdot (0.98)^t$, where t is the number of years since 2005.

   a. About how many lions were there in 2005?

   b. Describe what is happening to the population of lions over this decade.

   c. About how many lions are there in 2015? Show your reasoning.

NAME _____ DATE _____ PERIOD _____

4. A bank account pays 0.5% monthly interest.

   a. If $500 is put in the account, what will the balance be at the end of one year, assuming no additional deposits or withdrawals are made?

   b. What is the effective annual interest rate?

   c. Is the effective annual interest rate more or less than 6% (the nominal interest rate)?

5. Here are the graphs of three equations: $y = 50 \cdot (1.5)^x$, $y = 50 \cdot 2^x$, and $y = 50 \cdot (2.5)^x$.

   Which equation matches each graph? Explain how you know. (Lesson 5-12)

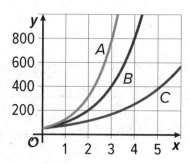

6. A major retailer has a staff of 6,400 employees for the holidays. After the holidays, they will decrease their staff by 30%.

   How many employees will they have after the holidays? (Lesson 5-14)

7. Ten students guessed the number of cubes in a jar that contains 202 cubes. Their names and guesses are listed in the table.

   Create a scatter plot with the guesses as the horizontal values and the absolute guessing errors as the vertical values. (Lesson 4-13)

Andre	205
Clare	190
Diego	197
Elena	200
Han	220
Jada	210
Kiran	202
Lin	203
Mai	199
Noah	185

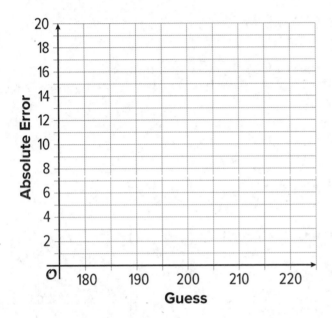

Lesson 5-17

Different Compounding Intervals

NAME _____ DATE _____ PERIOD _____

Learning Goal Let's find out what happens when we repeatedly apply the same percent increase at different intervals of time.

 ## Warm Up
17.1 Returns Over Three Years

Earlier, you learned about a bank account that had an initial balance of $1,000 and earned 1% monthly interest. Each month, the interest was added to the account and no other deposits or withdrawals were made.

To calculate the account balance in dollars after 3 years, Elena wrote: $1{,}000 \cdot (1.01)^{36}$ and Tyler wrote: $1{,}000 \cdot ((1.01)^{12})^3$.

Discuss with a partner:

1. Why do Elena's expression and Tyler's expression both represent the account balance correctly?

2. Kiran said, "The account balance is about $1{,}000 \cdot (1.1268)^3$." Do you agree? Why or why not?

Activity

17.2 Contemplating Credit Cards

A credit card company lists a nominal APR (annual percentage rate) of 24% but compounds interest monthly, so it calculates 2% per month.

Suppose a cardholder made $1,000 worth of purchases using his credit card and made no payments or other purchases. Assume the credit card company does not charge any additional fees other than the interest.

1. Write expressions for the balance on the card after 1 month, 2 months, 6 months, and 1 year.

2. Write an expression for the balance on the card, in dollars, after m months without payment.

3. How much does the cardholder owe after 1 year without payment? What is the *effective* APR of this credit card?

4. Write an expression for the balance on the card, in dollars, after t years without payment. Be prepared to explain your expression.

NAME _____ DATE _____ PERIOD _____

Are you ready for more?

A bank account has an annual interest rate of 12% and an initial balance of $800. Any earned interest is added to the account, but no other deposits or withdrawals are made. Write an expression for the account balance:

1. After 5 years, if interest is compounded n times per year.

2. After t years, if interest is compounded n times per year.

3. After t years, with an initial deposit of P dollars and an annual interest percentage rate of r, compounded n times per year.

Activity

17.3 Which One Would You Choose?

Suppose you have $500 to invest and can choose between two investment options.

- Option 1: every 3 months 3% interest is applied to the balance

- Option 2: every 4 months 4% interest is applied to the balance

Which option would you choose? Build a mathematical model for each investment option and use them to support your investment decision. Remember to state your assumptions about the situation.

Is there a period of time during which the first option (3% interest rate, compounded quarterly) will *always* be the better option? If so, when might it be? If not, why might that be?

NAME _____ DATE _____ PERIOD _____

Activity
17.4 Changes Over the Years

1. The function f defined by $f(x) = 15 \cdot (1.07)^x$ models the cost of tuition, in thousands of dollars, at a local college x years since 2017.

   a. What is the cost of tuition at the college in 2017?

   b. At what annual percentage rate does the tuition grow?

   c. Assume that before 2017 the tuition had also been growing at the same rate as after 2017. What was the tuition in 2000? Show your reasoning.

   d. What was the tuition in 2010?

   e. What will the tuition be when you graduate from high school?

2. Between 2000 and 2010 the tuition nearly doubled.

   a. By what factor will the tuition grow between 2017 and 2027? Show your reasoning.

   b. Choose another 10-year period and find the factor by which the tuition grows. Show your reasoning.

   c. What can you say about how the tuition changes over any 10-year period (assuming the function f continues to be an accurate model)? Explain or show how you know that this will *always* be the case.

Summary
Different Compounding Intervals

In many situations, interest is calculated more frequently than once a year. How often the interest is compounded and calculated and added to the previous amount affects the overall amount of interest earned (or owed) over time.

Suppose a bank account has a balance of $1,000 and a nominal annual interest rate of 6% per year. No additional deposits or withdrawals are made.

If the bank compounds interest annually, the account will have one interest calculation in one year, at a 6% rate. If it compounds interest every 6 months, the account will see two interest calculations in one year, at a 3% rate each time (because $6 \div 2 = 3$). If it is compounded every 3 months, there will be 4 calculations at 1.5% each time, and so on.

This table shows the nominal interest rates used for different compounding intervals, as well as the corresponding expressions for the account balance in one year.

Compounding Interval	Compounding Frequency per Year	Nominal Interest Rate	Account Balance in One Year
annually (12 months)	1 time	6%	$1,000 \cdot (1 + 0.06)$
semi-annually (6 months)	2 times	3%	$1,000 \cdot (1 + 0.03)^2$
quarterly (3 months)	4 times	1.5%	$1,000 \cdot (1 + 0.015)^4$
monthly (1 month)	12 times	0.5%	$1,000 \cdot (1 + 0.005)^{12}$

If we evaluate the expressions, we find these account balances:

- annually: $1,000 \cdot (1.06) = 1,060$
- semi-annually: $1,000 \cdot (1.03)^2 = 1,060.90$
- quarterly: $1,000 \cdot (1.015)^4 \approx 1,061.36$
- monthly: $1,000 \cdot (1.005)^{12} \approx 1,061.68$

Notice that the more frequently interest is calculated, the greater the balance is.

NAME _____ DATE _____ PERIOD _____

 Practice
Different Compounding Intervals

1. The population of a city in 2010 is 50,000, and it grows by 5% each year after.

   a. Write a function f which models the population of the city t years after 2010.

   b. What is the population of the city in 2017?

   c. What will the population of the city be in 2020? What about in 2030?

   d. By what factor does the population grow between 2010 and 2020? What about between 2020 and 2030?

2. A person charges $100 to a credit card with a 24% nominal annual interest rate.

   Assuming no other charges or payments are made, find the balance on the card, in dollars, after 1 year if interest is calculated:

   a. annually

   b. every 6 months

   c. every 3 months

   d. monthly

   e. daily

3. A couple has $5,000 to invest and has to choose between three investment options.

- Option A: $2\frac{1}{4}$% interest applied each quarter
- Option B: 3% interest applied every 4 months
- Option C: $4\frac{1}{2}$% interest applied twice a year

If they plan on no deposits and no withdrawals for 5 years, which option will give them the largest balance after 5 years? Use a mathematical model for each option to explain your choice.

4. Elena says that 6% interest applied semi-annually is the same as 1% interest applied every month: she reasons they are the same because they are both a 12% nominal annual interest rate.

   a. Is Elena correct that these two situations both offer a 12% nominal annual interest rate?

   b. Is Elena correct that the two situations pay the same amount of interest?

NAME _____ DATE _____ PERIOD _____

5. A bank pays 8% nominal annual interest, compounded at the end of each month. An account starts with $600, and no further withdrawals or deposits are made.

   a. What is the monthly interest rate?

   b. Write an expression for the account balance, in dollars, after one year.

   c. What is the effective annual interest rate?

   d. Write an expression for the account balance, in dollars, after t years.

6. At the end of each year, 10% interest is charged on a $500 loan. The interest applies to any unpaid balance on the loan, including previous interest.

   Select **all** the expressions that represent the loan balance after two years if no payments are made. **(Lesson 5-15)**

   A. $500 + 2 \cdot (0.1) \cdot 500$

   B. $500 \cdot (1.1) \cdot (1.1)$

   C. $500 + (0.1) + (0.1)$

   D. $500 \cdot (1.1)^2$

   E. $(500 + 50) \cdot (1.1)$

7. Here is a graph of the function f given by $f(x) = 100 \cdot 2^x$.

   Suppose g is the function given by $g(x) = 50 \cdot (1.5)^x$.

   Will the graph of g meet the graph of f for any positive value of x? Explain how you know. (Lesson 5-12)

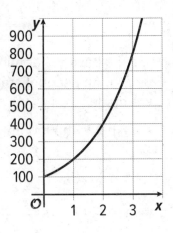

8. Suppose m and c each represent the position number of a letter in the alphabet, but m represents the letters in the original message and c the letters in a secret code.

   The equation $c = m + 7$ is used to encode a message. (Lesson 4-15)

   a. Write an equation that can be used to decode the secret code into the original message.

   b. What does this code say: "AOPZ PZ AYPJRF!"?

Lesson 5-18

Expressed in Different Ways

NAME _____ DATE _____ PERIOD _____

Learning Goal Let's write exponential expressions in different ways.

 Warm Up

18.1 Math Talk: Equal Expressions

Decide if each expression is equal to $(1.21)^{100}$.

$\left((1.21)^{10}\right)^{10}$

$\left((1.21)^{50}\right)^{50}$

$\left((1.1)^{2}\right)^{100}$

$(1.1)^{200}$

Activity

18.2 Population Projections

1. From 1790 to 1860, the United States population, in thousands, is modeled by the equation $P = 4{,}000 \cdot (1.031)^t$ where t is the number of years since 1790.

   a. About how many people were living in the U.S. in 1790? What about in 1860? Show your reasoning.

   b. What is the approximate annual percent increase predicted by the model?

   c. What does the model predict for the population in 2017? Is it accurate? Explain.

NAME _____ DATE _____ PERIOD _____

2. Respond to each question.

   a. What percent increase does the model predict each decade? Explain.

   b. Suppose *d* represents the number of *decades* since 1790. Write an equation for *P* in terms of *d* modeling the population in the US (in thousands).

3. Respond to each question.

   a. What percent increase does the model predict each century? Explain.

   b. Suppose *c* represents the number of centuries since 1790. Write an equation for *P* in terms of *c* modeling the population in the United States (in thousands).

Here are three expressions and three descriptions. In each case, $1,000 has been put in an interest-bearing bank account. No withdrawals or other deposits (aside from the earned interest) are made for 6 years.

- $1,000 \cdot \left(1 + \dfrac{0.07}{12}\right)^{72}$

- $1,000 \cdot \left(1 + \dfrac{0.07}{2}\right)^{12}$

- $1,000 \cdot \left(\left(1 + \dfrac{0.07}{12}\right)^{12}\right)^{6}$

- 7% annual interest compounded semi-annually

- 7% annual interest compounded monthly

- 7% annual interest compounded every two months

Sort the expressions and descriptions that represent the same amounts of interest into groups. One group contains more than two expressions. One of the descriptions does not have a match. Write an expression that matches it.

NAME _____ DATE _____ PERIOD _____

Are you ready for more?

Investing $1,000 at a 5% annual interest rate for 6 years, compounded every two months, yields $1,348.18. Without doing any calculations, rank these four possible changes in order of the increase in the interest they would yield from the greatest increase to the least increase:

- Increase the starting amount by $100.

- Increase the interest rate by 1%.

- Let the account increase for one more year.

- Compound the interest every month instead of every two months.

Once you have made your predictions, calculate the value of each option to see if your ranking was correct.

Expressions can be written in different ways to highlight different aspects of a situation or to help us better understand what is happening. A **growth rate** tells us the percent change. As always, in percent change situations, it is important to know if the change is an increase or decrease. For example:

- A population is increasing by 20% each year. The growth rate is 20%, so after one year, 0.2 times the population at the beginning of that year is being added. If the initial population is p, the new population is $p + 0.2p$, which equals $(1 + 0.2)p$ or $1.2p$.

- A population is decreasing by 20% each year. The growth rate is -20%, so after one year, 0.2 times the population at the beginning of that year is being lost. If the initial population is p, the new population is $p - 0.2p$, which equals $(1 - 0.2)p$ or $0.8p$.

Suppose the area a covered by a forest is currently 50 square miles and it is growing by 0.2% each year. If t represents time, from now, in years we can express the area of the forest as:

$$a = 50 \cdot (1 + 0.002)^t$$

$$a = 50 \cdot (1.002)^t$$

In this situation, the *growth rate* is 0.002, and the growth factor is 1.002. Because 0.002 is such a small number, however, it may be difficult to tell from this function how quickly the forest is growing. We may find it more meaningful to measure the growth every decade or every century. There are 10 years in a decade, so to find the growth rate in decades, we can use the expression $(1.002)^{10}$, which is approximately 1.02. This means a growth rate of about 2% per decade. Using d for time, in decades, the area of the forest can be expressed as:

$$a = 50 \cdot ((1 + 0.002)^{10})^d$$

$$a \approx 50 \cdot (1.02)^d$$

If we measure time in centuries, the growth rate is about 22% per century because $1.002^{100} \approx 1.22$. Using c to measure time, in centuries, our equation for area becomes:

$$a = 50 \cdot ((1 + 0.002)^{100})^c$$

$$a \approx 50 \cdot (1.22)^c$$

Glossary

growth rate

NAME _____ DATE _____ PERIOD _____

 Practice

Expressed in Different Ways

1. For each growth rate, find the associated growth factor.

  a. 30% increase

  b. 30% decrease

  c. 2% increase

  d. 2% decrease

  e. 0.04% increase

  f. 0.04% decrease

  g. 100% increase

2. In 1990, the population p of India was about 870.5 million people. By 1995, there were about 960.9 million people. The equation $p = 870.5 \cdot (1.021)^t$ approximates the number of people, in millions, in terms of the number of years t since 1990.

  a. By what factor does the number of people grow in one year?

  b. If d is time in decades, write an equation expressing the number of people in millions, p, in terms of decades, d, since 1990.

  c. Use the model $p = 870.5 \cdot (1.021)^t$ to predict the number of people in India in 2015.

  d. In 2015, the population of India was 1,311 million. How does this compare with the predicted number?

3. An investor paid $156,000 for a condominium in Texas in 2008. The value of the homes in the neighborhood have been appreciating by about 12% annually.

Select **all** the expressions that could be used to calculate the value of the house, in dollars, after t years.

(A.) $156{,}000 \cdot (0.12)^t$

(B.) $156{,}000 \cdot (1.12)^t$

(C.) $156{,}000 \cdot (1 + 0.12)^t$

(D.) $156{,}000 \cdot (1 - 0.12)^t$

(E.) $156{,}000 \cdot \left(1 + \frac{0.12}{12}\right)^t$

4. A credit card has a nominal annual interest rate of 18%, and interest is compounded monthly. The cardholder uses the card to make a $30 purchase.

Which expression represents the balance on the card after 5 years, in dollars, assuming no further charges or payments are made?

(A.) $30(1 + 18)^5$

(B.) $30(1 + 0.18)^5$

(C.) $30\left(1 + \frac{0.18}{12}\right)^5$

(D.) $30\left(1 + \frac{0.18}{12}\right)^{5 \cdot 12}$

NAME _____ DATE _____ PERIOD _____

5. The expression $1{,}500 \cdot (1.085)^3$ represents an account balance in dollars after three years with an initial deposit of $1,500. The account pays 8.5% interest, compounded annually for three years.

    a. Explain how the expression would change if the bank had compounded the interest quarterly for the three years.

    b. Write a new expression to represent the account balance, in dollars, if interest is compounded quarterly.

6. The function, f, defined by $f(t) = 1{,}000 \cdot (1.07)^t$, represents the amount of money in a bank account t years after it was opened. **(Lesson 5-9)**

    a. How much money was in the account when it was opened?

    b. Sketch a graph of f

    c. When does the account value reach $2,000?

7. The graph shows the number of patients with an infectious disease over a period of 15 weeks. (Lesson 5-10)

a. Give an example of a domain for which the average rate of change is a good measure of how the function changes.

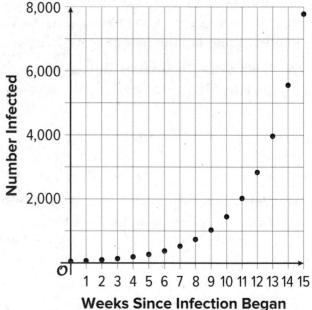

b. Give an example of a domain for which the average rate of change is not a good measure of how the function changes.

8. A party will have pentagonal tables placed together. The number of people, P, who can sit at the tables is a function of the number of tables, n.
(Lesson 4-16)

a. Explain why the equation $P = 3n + 2$ defines this function.

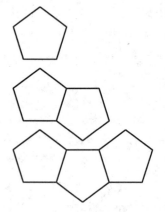

b. How many tables are needed if 47 people come to the party?

c. How many tables are needed if 99 people come to the party?

d. Write the inverse of this function and explain what the inverse function tells us.

Lesson 5-19
Which One Changes Faster?

NAME _____ DATE _____ PERIOD _____

Learning Goal Let's compare linear and exponential functions as they continue to increase.

Warm Up
19.1 Graph of Which Function?

Here is a graph.

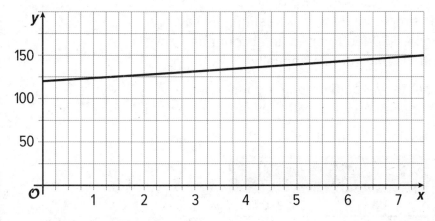

1. Which equation do you think the graph represents? Use the graph to support your reasoning.

    • $y = 120 + (3.7) \cdot x$

    • $y = 120 \cdot (1.03)^x$

2. What information might help you decide more easily whether the graph represents a linear or an exponential function?

 Activity

19.2 Simple and Compound Interests

A family has $1,000 to invest and is considering two options: investing in government bonds that offer 2% simple interest, or investing in a savings account at a bank, which charges a $20 fee to open an account and pays 2% compound interest. Both options pay interest annually.

Here are two tables showing what they would earn in the first couple of years if they do not invest additional amounts or withdraw any money.

Bonds

Years of Investment	Amount in Dollars
0	$1,000
1	$1,020
2	$1,040

Savings Account

Years of Investment	Amount in Dollars
0	$980
1	$999.60
2	$1,019.59

1. Bonds: How does the investment grow with simple interest?

2. Savings account: How are the amounts $999.60 and $1,019.59 calculated?

3. For each option, write an equation to represent the relationship between the amount of money and the number of years of investment.

NAME _____ DATE _____ PERIOD _____

4. Which investment option should the family choose? Use your equations or calculations to support your answer.

5. Use graphing technology to graph the two investment options and show how the money grows in each.

Activity
19.3 Reaching 2,000

1. Complete the table of values for the functions *f* and *g*.

x	f(x)	g(x)
1		
10		
50		
100		
500		

2. Based on the table of values, which function do you think grows faster? Explain your reasoning.

3. Which function do you think will reach a value of 2,000 first? Show your reasoning. If you get stuck, consider increasing x by 100 a few times and record the function values in the table.

Are you ready for more?

Consider the functions $g(x) = x^5$ and $f(x) = 5^x$. While it is true that $f(7) > g(7)$, for example, it is hard to check this using mental math. Find a value of x for which properties of exponents allow you to conclude that $f(x) > g(x)$ without a calculator.

NAME _____ DATE _____ PERIOD _____

Summary
Which One Changes Faster?

Suppose that you won the top prize from a game show and are given two options. The first option is a cash gift of $10,000 and $1,000 per day for the next 7 days. The second option is a cash gift of 1 cent (or $0.01) that grows tenfold each day for 7 days. Which option would you choose?

In the first option, the amount of money increases by the same amount ($1,000) each day, so we can represent it with a linear function. In the second option, the money grows by multiples of 10, so we can represent it with an exponential function. Let f represent the amount of money x days after winning with the first option and let g represent the amount of money x days after winning with the second option.

Option 1: $f(x) = 10,000 + 1,000x$ Option 2: $g(x) = (0.01) \cdot 10^x$

$f(1) = 11,000$ $g(1) = 0.1$

$f(2) = 12,000$ $g(2) = 1$

$f(3) = 13,000$ $g(3) = 10$

$. . .$ $. . .$

$f(6) = 16,000$ $g(6) = 10,000$

$f(7) = 17,000$ $g(7) = 100,000$

For the first few days, the second option trails far behind the first. Because of the repeated multiplication by 10, however, after 7 days it surges past the amount in the first option.

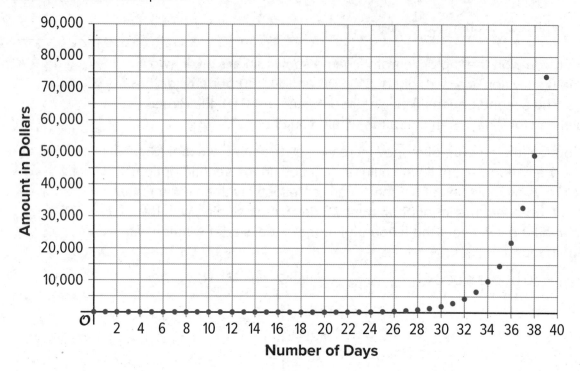

What if the factor of growth is much smaller than 10? Suppose we have a third option, represented by a function h. The starting amount is still $0.01 and it grows by a factor of 1.5 times each day. If we graph the function $h(x) = (0.01) \cdot (1.5)^x$, we see that it takes many, many more days before we see a rapid growth. But given time to continue growing, the amount in this exponential option will eventually also outpace that in the linear option.

NAME _____ DATE _____ PERIOD _____

Practice
Which One Changes Faster?

1. Functions a, b, c, d, e, and f are given below. Classify each function as linear, exponential, or neither.

   a. $a(x) = 3x$

   b. $b(x) = 3^x$

   c. $c(x) = x^3$

   d. $d(x) = 9 + 3x$

   e. $e(x) = 9 \cdot 3^x$

   f. $f(x) = 9 \cdot 3x$

2. Here are 4 equations defining 4 different functions, a, b, c, and d. List them in order of increasing rate of change. That is, start with the one that grows the slowest and end with the one that grows the quickest.

   $a(x) = 5x + 3 \qquad b(x) = 3x + 5 \qquad c(x) = x + 4 \qquad d(x) = 1 + 4x$

3. *Technology required.* Function f is defined by $f(x) = 3x + 5$ and function g is defined by $g(x) = (1.1)^x$.

a. Complete the table with values of $f(x)$ and $g(x)$. When necessary, round to 2 decimal places.

x	f(x)	g(x)
1		
5		
10		
20		

b. Which function do you think grows faster? Explain your reasoning.

c. Use technology to create graphs representing f and g. What graphing window do you have to use to see the value of x where g becomes greater than f for that x?

4. Functions m and n are given by $m(x) = (1.05)^x$ and $n(x) = \frac{5}{8}x$. As x increases from 0:

a. Which function reaches 30 first?

b. Which function reaches 100 first?

5. The functions f and g are defined by $f(x) = 8x + 33$ and $g(x) = 2 \cdot (1.2)^x$.

a. Which function eventually grows faster, f or g? Explain how you know.

b. Explain why the graphs of f and g meet for a positive value of x.

NAME _____ DATE _____ PERIOD _____

6. A line segment of length ℓ is scaled by a factor of 1.5 to produce a segment with length m. The new segment is then scaled by a factor of 1.5 to give a segment of length n.

   What scale factor takes the segment of length ℓ to the segment of length n? Explain your reasoning. (Lesson 5-16)

7. A couple needs to get a loan of $5,000 and has to choose between three options.

   - Option A: $2\frac{1}{4}$% applied quarterly

   - Option B: 3% applied every 4 months

   - Option C: $4\frac{1}{2}$% applied semi-annually

   If they make no payments for 5 years, which option will give them the least amount owed after 5 years? Use a mathematical model for each option to explain your choice. (Lesson 5-17)

8. Here are graphs of five absolute value functions. Match the graph and equation that represent the same function. **(Lesson 4-14)**

Graph 1

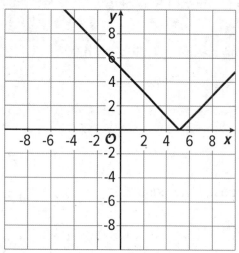

Graph 2

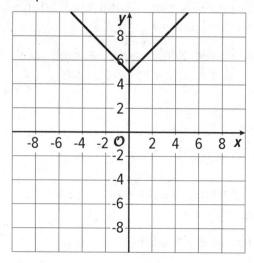

Graph 3

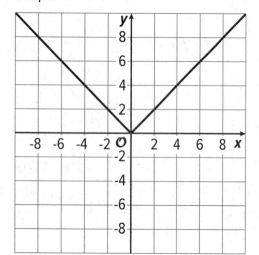

Graph 4

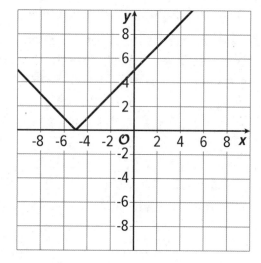

A. $f(x) = |x|$

B. $f(x) = |x - 5|$

C. $f(x) = |x| - 5$

D. $f(x) = |x + 5|$

E. $f(x) = |x| + 5$

1. Graph 1

2. Graph 2

3. Graph 3

4. Graph 4

5. Graph 5

Graph 5

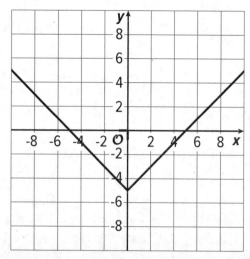

Lesson 5-20

Changes over Equal Intervals

DATE _____ PERIOD _____

NAME _____

Learning Goal Let's explore how linear and exponential functions change over equal intervals.

Warm Up
20.1 Writing Equivalent Expressions

For each given expression, write an equivalent expression with as few terms as possible.

1. $7p - 3 + 2(p + 1)$

2. $[4(n + 1) + 10] - 4(n + 1)$

3. $9^5 \cdot 9^2 \cdot 9^x$

4. $\dfrac{2^{4n}}{2^n}$

Activity
20.2 Outputs of a Linear Function

Here is a graph of $y = f(x)$ where $f(x) = 2x + 5$.

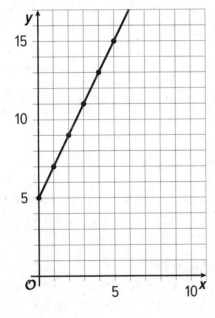

1. How do the values of f change whenever x increases by 1, for instance, when it increases from 1 to 2, or from 19 to 20? Be prepared to explain or show how you know.

2. Here is an expression we can use to find the difference in the values of f when the input changes from x to x + 1.

$$[2(x + 1) + 5] - [2x + 5]$$

   Does this expression have the same value as what you found in the previous questions? Show your reasoning.

3. Respond to each question.

   a. How do the values of f change whenever x increases by 4? Explain or show how you know.

   b. Write an expression that shows the change in the values of f when the input value changes from x to x + 4.

   c. Show or explain how that expression has a value of 8.

NAME _____ DATE _____ PERIOD _____

Activity

20.3 Outputs of an Exponential Function

Here is a table that shows some input and output values of an exponential function g. The equation $g(x) = 3^x$ defines the function.

x	$g(x)$
3	27
4	81
5	243
6	729
7	2,187
8	6,561
x	
$x + 1$	

1. How does $g(x)$ change every time x increases by 1? Show or explain your reasoning.

2. Choose two new input values that are consecutive whole numbers and find their output values. Record them in the table. How do the output values change for those two input values?

3. Complete the table with the output when the input is x and when it is $x + 1$.

4. Look at the change in output values as the x increases by 1. Does it still agree with your findings earlier? Show your reasoning.

   Pause here for a class discussion. Then, work with your group on the next few questions.

5. Choose two x-values where one is 3 more than the other (for example, 1 and 4). How do the output values of g change as x increases by 3? (Each group member should choose a different pair of numbers and study the outputs.)

6. Complete this table with the output when the input is x and when it is $x + 3$. Look at the change in output values as x increases by 3. Does it agree with your group's findings in the previous question? Show your reasoning.

x	$g(x)$
x	
$x + 3$	

For integer inputs, we can think of multiplication as repeated addition and exponentiation as repeated multiplication:

$$3 \cdot 5 = 3 + 3 + 3 + 3 + 3 \quad \text{and} \quad 3^5 = 3 \cdot 3 \cdot 3 \cdot 3 \cdot 3$$

We could continue this process with a new operation called tetration. It uses the symbol ↑↑, and is defined as repeated exponentiation:

$$3 \uparrow\uparrow 5 = 3^{3^{3^{3^{3}}}}.$$

Compute $2 \uparrow\uparrow 3$ and $3 \uparrow\uparrow 2$. If $f(x) = 3 \uparrow\uparrow x$, what is the relationship between $f(x)$ and $f(x + 1)$?

NAME _____ DATE _____ PERIOD _____

Summary
Changes over Equal Intervals

Linear and exponential functions each behave in a particular way every time their input value increases by the same amount.

Take the linear function f defined by $f(x) = 5x + 3$. The graph of this function has a slope of 5. That means that each time x increases by 1, $f(x)$ increases by 5. For example, the points (7, 38) and (8, 43) are both on the graph. When x increases by 1 (from 7 to 8), y increases by 5 (because $43 - 38 = 5$). We can show algebraically that this is always true, regardless of what value x takes.

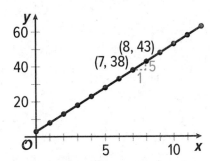

The value of f when x increases by 1, or $f(x + 1)$, is $5(x + 1) + 3$. Subtracting $f(x + 1)$ and $f(x)$, we have:

$$f(x + 1) - f(x) = 5(x + 1) + 3 - (5x + 3)$$
$$= 5x + 5 + 3 - 5x - 3$$
$$= 5$$

This tells us that whenever x increases by 1, the difference in the output is always 5. In the lesson, we also saw that when x increases by an amount other than 1, the output always increases by the same amount if the function is linear.

Now let's look at an exponential function g defined by $g(x) = 2^x$. If we graph g, we see that each time x increases by 1, the value $g(x)$ doubles. We can show algebraically that this is always true, regardless of what value x takes.

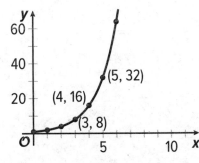

The value of g when x increases by 1, or $g(x + 1)$, is 2^{x+1}. Dividing $g(x + 1)$ by $g(x)$, we have:

$$\frac{g(x + 1)}{g(x)} = \frac{2^{(x+1)}}{2^x}$$
$$= 2^{x+1-x}$$
$$= 2^1$$
$$= 2$$

This means that, whenever x increases by 1, the value of g always increases by a multiple of 2. In the lesson, we also saw that when x increases by an amount other than 1, the output always increases by the same factor if the function is exponential.

A linear function always increases (or decreases) by the same amount over equal intervals. An exponential function increases (or decreases) by equal *factors* over equal intervals.

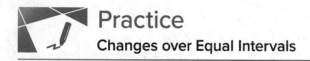
1. Whenever the input of a function f increases by 1, the output increases by 5. Which of these equations could define f?

   A. $f(x) = 3x + 5$

   B. $f(x) = 5x + 3$

   C. $f(x) = 5^x$

   D. $f(x) = x^5$

2. The function f is defined by $f(x) = 2^x$. Which of the following statements is true about the values of f? Select **all** that apply.

   A. When the input x increases by 1, the value of f increases by 2.

   B. When the input x increases by 1, the value of f increases by a factor of 2.

   C. When the input x increases by 3, the value of f increases by 8.

   D. When the input x increases by 3, the value of f increases by a factor of 8.

   E. When the input x increases by 4, the value of f increases by a factor of 4.

3. The two lines on the coordinate plane are graphs of functions f and g.

   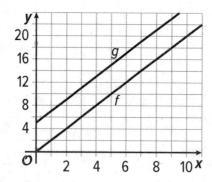

   a. Use the graph to explain why the value of f increases by 2 each time the input x increases by 1.

NAME _____ DATE _____ PERIOD _____

b. Use the graph to explain why the value of g increases by 2 each time the input x increases by 1.

4. The function h is given by $h(x) = 5^x$.

a. Find the quotient $\dfrac{h(x+2)}{h(x)}$.

b. What does this tell you about how the value of h changes when the input is increased by 2?

c. Find the quotient $\dfrac{h(x+3)}{h(x)}$.

d. What does this tell you about how the value of h changes when the input is increased by 3?

5. For each of the functions $f, g, h, p,$ and q, the domain is $0 \le x \le 100$. For which functions is the average rate of change a good measure of how the function changes for this domain? Select **all** that apply. (Lesson 5-10)

(A.) $f(x) = x + 2$

(B.) $g(x) = 2^x$

(C.) $h(x) = 111x - 23$

(D.) $p(x) = 50{,}000 \cdot 3^x$

(E.) $q(x) = 87.5$

6. The average price of a gallon of regular gasoline in 2016 was $2.14. In 2017, the average price was $2.42 a gallon—an increase of 13%.

   At that rate, what will the average price of gasoline be in 2020? (Lesson 5-16)

7. A credit card charges a 14% annual nominal interest rate and has a balance of $500.

   If no payments are made and interest is compounded quarterly, which expression could be used to calculate the account balance, in dollars, in 3 years? (Lesson 5-17)

   (A.) $500 \cdot (1 + 0.14)^3$

   (B.) $500 \cdot \left(1 + \frac{0.14}{4}\right)^3$

   (C.) $500 \cdot \left(1 + \frac{0.14}{4}\right)^{12}$

   (D.) $500 \cdot \left(1 + \frac{0.14}{4}\right)^{48}$

8. Here are equations that define four linear functions. For each function, write a verbal description of what is done to the input to get the output, and then write the inverse function. (Lesson 4-17)

   a. $a(x) = x - 4$

   b. $b(x) = 2x - 4$

   c. $c(x) = 2(x - 4)$

   d. $d(x) = \frac{x}{4}$

Lesson 5-21

Predicting Populations

NAME _____ DATE _____ PERIOD _____

Learning Goal Let's use linear and exponential models to represent and understand population changes.

 ## Warm Up
21.1 Notice and Wonder: Changing Populations

Here are the populations of three cities during different years.

City	1950	1960	1970	1980	1990	2000
Paris	6,300,000	7,400,000	8,200,000	8,700,000	9,300,000	9,700,000
Austin	132,000	187,000	254,000	346,000	466,000	657,000
Chicago	3,600,000	3,550,000	3,400,000	3,000,000	2,800,000	2,900,000

What do you notice? What do you wonder?

Here are population data for three cities at different times between 1950 and 2000. What does the data tell us, if anything, about the current population in the cities or what the population will be in 2050?

City	1950	1960	1970	1980	1990	2000
Paris	6,300,000	7,400,000	8,200,000	8,700,000	9,300,000	9,700,000
Austin	132,000	187,000	254,000	346,000	466,000	657,000
Chicago	3,600,000	3,550,000	3,400,000	3,000,000	2,800,000	2,900,000

1. How would you describe the population change in each city during this time period? Write one to two sentences for each city. Then discuss with your group.

NAME _____ DATE _____ PERIOD _____

2. What kind of model, linear or exponential, both, or neither do you think is appropriate for each city population?

3. For each population that you think can be modeled by a linear and or exponential function:

   a. Write an equation for the function(s).

   b. Graph the function(s).

NAME _____ DATE _____ PERIOD _____

4. Compare the graphs of your functions with the actual population data. How well do the models fit the data?

5. Respond to each question.

   a. Use your models to predict the population in each city in 2010, the current year, and 2050.

   b. Do you think that these predictions are (or will be) accurate? Explain your reasoning.

Year	1804	1927	1960	1974	1987	1999	2011
World Population in Billions	1	2	3	4	5	6	7

1. Would a linear function be appropriate for modeling the world population growth over the last 200 years? Explain. If you think it is appropriate, find a linear model.

2. Would an exponential function be appropriate for modeling the world population growth over the last 200 years? Explain. If you think it is appropriate, find an exponential model.

NAME _____ DATE _____ PERIOD _____

3. From 1950 to the present day, by about what percentage has the world population grown each year?

4. From 1950 to the present day, by about how many people has the world population grown each year?

5. If the growth trend continues, what will the world population be in 2050? How long do you think the growth will continue? Explain your reasoning.

Are you ready for more?

Another common model for population growth which fixes some of the improbable predictions of the exponential model is called a *logistic* model. A sample function f of this type is the function

$$f(t) = \frac{10}{1 + 50 \cdot 2^{-t}}.$$

Evaluate this function for integer values of t between 0 and 15. Describe qualitatively how this function differs from an exponential one. What happens to the world population in the long run according to this model?

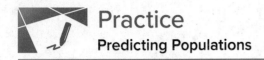

Practice

Predicting Populations

1. The table shows the height of a ball after different numbers of bounces. (Lesson 5-11)

n	1	2	3	4	5
h	83	61	46	35	26

a. Can the height, h, in centimeters, after n bounces be modeled accurately by a linear function? Explain your reasoning.

b. Can the height, h, after n bounces be modeled accurately by an exponential function? Explain your reasoning

c. Create a model for the height of the ball after n bounces and plot the predicted values with the data.

d. Use your model to estimate the height the ball was dropped from.

e. Use your model to estimate how many bounces it takes before the rebound height is less than 10 cm.

NAME _____ DATE _____ PERIOD _____

2. Mai used a computer simulation to roll number cubes and count how many rolls it took before all of the cubes came up sixes. Here is a table showing her results. **(Lesson 5-11)**

d, Number of Cubes	1	2	3	4
r, Number of Rolls	5	31	143	788

Would a linear or exponential function be appropriate for modeling the relationship between d and r? Explain how you know.

3. A ramp is two meters long. Priya wants to investigate how the distance a basketball rolls is related to the location on the ramp where it is released. **(Lesson 5-11)**

Recommend a way Priya can gather data to help understand this relationship.

4. Here are the graphs of three functions. **(Lesson 5-12)**

Which of these functions decays the most quickly? Which one decays the least quickly?

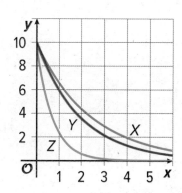

5. The bungee jump in Rishikesh, India is 83 meters high. The jumper free falls for 5 seconds to about 30 meters above the river. (Lesson 4-8)

a. Draw a graph of the bungee jump in Rishikesh.

b. Identify and describe three pieces of important information you can learn from the graph of the bungee jump.

Learning Targets

Lesson	Learning Target(s)
5-1 Growing and Growing	• I can compare growth patterns using calculations and graphs.
5-2 Patterns of Growth	• I can use words and expressions to describe patterns in tables of values. • When I have descriptions of linear and exponential relationships, I can write expressions and create tables of values to represent them.
5-3 Representing Exponential Growth	• I can explain the connections between an equation and a graph that represents exponential growth. • I can write and interpret an equation that represents exponential growth.
5-4 Understanding Decay	• I can use only multiplication to represent "decreasing a quantity by a fraction of itself." • I can write an expression or equation to represent a quantity that decays exponentially. • I know the meanings of "exponential growth" and "exponential decay."

(continued on the next page)

(continued from the previous page)

Lesson	Learning Target(s)
5-5 Representing Exponential Decay	• I can explain the meanings of a and b in an equation that represents exponential decay and is written as $y = a \cdot b^x$.
	• I can find a growth factor from a graph and write an equation to represent exponential decay.
	• I can graph equations that represent quantities that change by a growth factor between 0 and 1.
5-6 Analyzing Graphs	• I can use graphs to compare and contrast situations that involve exponential decay.
	• I can use information from a graph to write an equation that represents exponential decay.
5-7 Using Negative Exponents	• I can describe the meaning of a negative exponent in equations that represent exponential decay.
	• I can write and graph an equation that represents exponential decay to solve problems.
5-8 Exponential Situations as Functions	• I can use function notation to write equations that represent exponential relationships.
	• When I see relationships in descriptions, tables, equations, or graphs, I can determine whether the relationships are functions.

(continued on the next page)

(continued from the previous page)

Lesson	Learning Target(s)
5-9 Interpreting Exponential Functions	• I can analyze a situation and determine whether it makes sense to connect the points on the graph that represents the situation.
	• When I see a graph of an exponential function, I can make sense of and describe the relationship using function notation.
5-10 Looking at Rates of Change	• I can calculate the average rate of change of a function over a specified period of time.
	• I know how the average rate of change of an exponential function differs from that of a linear function.
5-11 Modeling Exponential Behavior	• I can use exponential functions to model situations that involve exponential growth or decay.
	• When given data, I can determine an appropriate model for the situation described by the data.
5-12 Reasoning about Exponential Graphs (Part 1)	• I can describe the effect of changing a and b on a graph that represents $f(x) = a \cdot b^x$.
	• I can use equations and graphs to compare exponential functions.

(continued on the next page)

(continued from the previous page)

Lesson	Learning Target(s)
5-13 Reasoning about Exponential Graphs (Part 2)	• I can explain the meaning of the intersection of the graphs of two functions in terms of the situations they represent. • When I know two points on a graph of an exponential function, I can write an equation for the function.
5-14 Recalling Percent Change	• I can find the result of applying a percent increase or decrease on a quantity. • I can write different expressions to represent a starting amount and a percent increase or decrease.
5-15 Functions Involving Percent Change	• I can use graphs to illustrate and compare different percent increases. • I can write a numerical expression or an algebraic expression to represent the result of applying a percent increase repeatedly.
5-16 Compounding Interest	• I can explain why applying a percent increase, p, n times is like or unlike applying the percent increase np.

(continued on the next page)

(continued from the previous page)

Lesson	Learning Target(s)
5-17 Different Compounding Intervals	• I can calculate interest when I know the starting balance, interest rate, and compounding intervals. • When given interest rates and compounding intervals, I can choose the better investment option.
5-18 Expressed in Different Ways	• I can solve problems using exponential expressions written in different ways. • I can write equivalent expressions to represent situations that involve repeated percent increase or decrease.
5-19 Which One Changes Faster?	• I can use tables, calculations, and graphs to compare growth rates of linear and exponential functions and predict how the quantities change eventually.
5-20 Changes over Equal Intervals	• I can calculate rates of change of functions given graphs, equations, or tables. • I can use rates of change to describe how a linear function and an exponential function change over equal intervals.

(continued on the next page)

(continued from the previous page)

Lesson	Learning Target(s)
5-21 Predicting Populations	• I can determine how well a chosen model fits the given information. • I can determine whether to use a linear function or an exponential function to model real-world data.

Notes:

Unit 6

Introduction to Quadratic Functions

The Italian physicist Galileo experimented with the speed of falling bodies in the 16th century. How long would it take an object to fall from this skyscraper? You will explore how to represent this relationship with quadratic functions in this unit.

Topics

- A Different Kind of Change
- Quadratic Functions
- Working with Quadratic Expressions
- Features of Graphs of Quadratic Functions

Unit 6

Introduction to Quadratic Functions

Lesson 6-1

A Different Kind of Change

NAME _____ DATE _____ PERIOD _____

Learning Goal Let's find the rectangle with the greatest area.

 ## Warm Up
1.1 Notice and Wonder: Three Tables

Look at the patterns in the 3 tables. What do you notice? What do you wonder?

x	y
1	0
2	5
3	10
4	15
5	20

x	y
1	3
2	6
3	12
4	24
5	48

x	y
1	8
2	11
3	10
4	5
5	-4

Activity

1.2 Measuring a Garden

Noah has 50 meters of fencing to completely enclose a rectangular garden in the backyard.

1. Draw some possible diagrams of Noah's garden. Label the length and width of each rectangle.

2. Find the length and width of such a rectangle that would produce the largest possible area. Explain or show why you think that pair of length and width gives the largest possible area.

Activity

1.3 Plotting the Measurements of the Garden

1. Plot some values for the length and area of the garden on the coordinate plane.

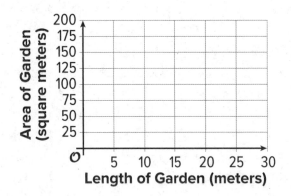

2. What do you notice about the plotted points?

NAME _____ DATE _____ PERIOD _____

3. The points (3, 66) and (22, 66) each represent the length and area of the garden. Plot these 2 points on the coordinate plane, if you haven't already done so. What do these points mean in this situation?

4. Could the point (1, 25) represent the length and area of the garden? Explain how you know.

Are you ready for more?

1. What happens to the area when you interchange the length and width? For example, compare the areas of a rectangle of length 11 meters and width 14 meters with a rectangle of length 14 meters and width 11 meters.

2. What patterns would you notice if you were to plot more length and area pairs on the graph?

In this lesson, we looked at the relationship between the side length and the area of a rectangle when the perimeter is unchanged.

If a rectangle has a perimeter of 40 inches, we can represent the possible lengths and widths as shown in the table.

Length (inches)	Width (inches)
2	18
5	15
10	10
12	8
15	5

We know that twice the length and twice the width must equal 40, which means that the length plus width must equal 20, or $\ell + w = 20$.

To find the width given a length ℓ, we can write: $w = 20 - \ell$.

The relationship between the length and the width is linear. If we plot the points from the table representing the length and the width, they form a line.

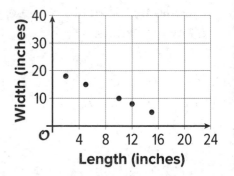

What about the relationship between the side lengths and the area of rectangles with perimeter of 40 inches?

Here are some possible areas of different rectangles whose perimeter are all 40 inches.

Length (inches)	Width (inches)	Area (square inches)
2	18	36
5	15	75
10	10	100
12	8	96
15	5	75

Here is a graph of the lengths and areas from the table:

Notice that, initially, as the length of the rectangle increases (for example, from 5 to 10 inches), the area also increases (from 75 to 100 square inches). Later, however, as the length increases (for example, from 12 to 15), the area decreases (from 96 to 75).

We have not studied relationships like this yet and will investigate them further in this unit.

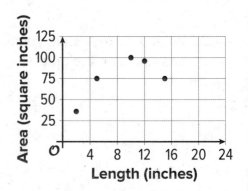

NAME _____ DATE _____ PERIOD _____

Practice
A Different Kind of Change

1. Here are a few pairs of positive numbers whose sum is 50.

   a. Find the product of each pair of numbers.

First Number	Second Number	Product
1	49	
2	48	
10	40	

   b. Find a pair of numbers that have a sum of 50 and will produce the largest possible product.

   c. Explain how you determined which pair of numbers have the largest product.

2. Here are some lengths and widths of a rectangle whose perimeter is 20 meters.

   a. Complete the table. What do you notice about the areas?

Length (meters)	Width (meters)	Area (square meters)
1	9	
3	7	
5		
7		
9		

   b. Without calculating, predict whether the area of the rectangle will be greater or less than 25 square meters if the length is 5.25 meters.

c. On the coordinate plane, plot the points for length and area from your table.

Do the values change in a linear way? Do they change in an exponential way?

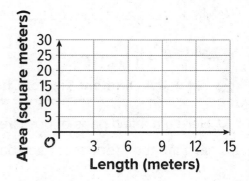

3. The table shows the relationship between x and y, the side lengths of a rectangle, and the area of the rectangle.

   a. Explain why the relationship between the side lengths is linear.

x (cm)	y (cm)	Area (sq cm)
2	4	8
4	8	32
6	12	72
8	16	128

   b. Explain why the relationship between x and the area is neither linear nor exponential.

4. Which statement best describes the relationship between a rectangle's side length and area as represented by the graph?

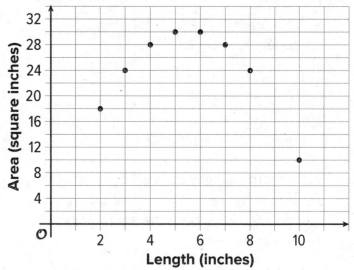

   (A.) As the side length increases by 1, the area increases and then decreases by an equal amount.

   (B.) As the side length increases by 1, the area increases and then decreases by an equal factor.

   (C.) As the side length increases by 1, the area does not increase or decrease by an equal amount.

   (D.) As the side length increases by 1, the area does not change.

NAME _____ DATE _____ PERIOD _____

5. Copies of a book are arranged in a stack. Each copy of a book is 2.1 cm thick. (Lesson 5-2)

   a. Complete the table.

Copies of Book	Stack Height in cm
0	
1	
2	
3	
4	

   b. What do you notice about the differences in the height of the stack of books when a new copy of the book is added?

   c. What do you notice about the factor by which the height of the stack of books changes when a new copy is added?

   d. How high is a stack of b books?

6. The value of a phone when it was purchased was $500. It loses $\frac{1}{5}$ of its value a year. (Lesson 5-4)

   a. What is the value of the phone after 1 year? What about after 2 years? 3 years?

   b. Tyler says that the value of the phone decreases by $100 each year since $\frac{1}{5}$ of 500 is 100. Do you agree with Tyler? Explain your reasoning.

7. *Technology required.* The data in the table represents the price of
   one gallon of milk in different years. (Lesson 3-5)

   Use graphing technology to create a
   scatter plot of the data.

   a. Does a linear model seem appropriate for this data?
      Why or why not?

   b. If it seems appropriate for the data, create the
      line of best fit. Round to two decimal places.

   c. What is the slope of the line of best fit, and
      what does it mean in this context? Is it realistic?

x, Time (years)	Price per Gallon of Milk (dollars)
1930	0.26
1935	0.47
1940	0.52
1940	0.50
1945	0.63
1950	0.83
1955	0.93
1960	1.00
1965	1.05
1970	1.32
1970	1.25
1975	1.57
1985	2.20
1995	2.50
2005	3.20
2018	2.90
2018	3.25

   d. What is the *y*-intercept of the line of best fit, and what does it mean in
      this context? Is it realistic?

8. Give a value for *r* that indicates that a line of best fit has a negative slope
   and models the data well. (Lesson 3-7)

Lesson 6-2

How Does it Change?

NAME _____ DATE _____ PERIOD _____

Learning Goal Let's describe some patterns of change.

 ## Warm Up
2.1 Squares in a Figure

How does each expression represent the number of small squares in the figure?

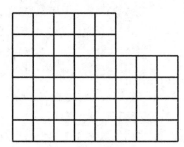

- Expression A: $6 \cdot 8 - 2 \cdot 3$

- Expression B: $4 \cdot 8 + 2 \cdot 5$

- Expression C: $8 + 8 + 8 + 8 + 5 + 5$

- Expression D: $5 \cdot 6 + 3 \cdot 4$

Pattern 1

Step 0   Step 1   Step 2   Step 3

Pattern 2

Step 0   Step 1   Step 2   Step 3

1. Study the 2 patterns of dots.

   a. How are the number of dots in each pattern changing?

   b. How would you find the number of dots in the 5th step in each pattern?

2. Complete the table with the number of dots in each pattern.

Step	Number of Dots in Pattern 1	Number of Dots in Pattern 2
0		
1		
2		
3		
4		
5		
10		
n		

NAME _____ DATE _____ PERIOD _____

3. Plot the number of dots at each step number.

Pattern 1

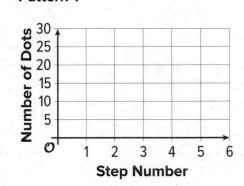

Pattern 2

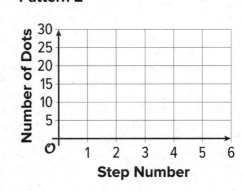

4. Explain why the graphs of the 2 patterns look the way they do.

Activity

2.3 Expressing a Growth Pattern

Step 1   **Step 2**   **Step 3**

Here is a pattern of squares.

1. Is the number of small squares growing linearly? Explain how you know.

2. Complete the table.

Step	Number of Small Squares
1	
2	
3	
4	
5	
10	
12	
n	

3. Is the number of small squares growing exponentially? Explain how you know.

Are you ready for more?

Han wrote $n(n + 2) - 2(n - 1)$ for the number of small squares in the design at Step n.

1. Explain why Han is correct.

2. Label the picture in a way that shows how Han saw the pattern when writing his expression.

NAME _____ DATE _____ PERIOD _____

Summary
How Does it Change?

In this lesson, we saw some quantities that change in a particular way, but the change is neither linear nor exponential. Here is a pattern of shapes, followed by a table showing the relationship between the step number and the number of small squares.

Step 1 **Step 2** **Step 3**

Step	Total Number of Small Squares
1	2
2	5
3	10
n	$n^2 + 1$

The number of small squares increases by 3, and then by 5, so we know that the growth is not linear. It is also not exponential because it is not changing by the same factor each time. From Step 1 to Step 2, the number of small squares grows by a factor of $\frac{5}{2}$, while from Step 2 to Step 3, it grows by a factor of 2.

From the diagram, we can see that in Step 2, there is a 2-by-2 square plus 1 small square added on top. Likewise, in Step 3, there is a 3-by-3 square with 1 small square added. We can reason that the nth step is an n-by-n arrangement of small squares with an additional small square on top, giving the expression $n^2 + 1$ for the number of small squares.

The relationship between the step number and the number of small squares is a quadratic relationship, because it is given by the expression $n^2 + 1$, which is an example of a **quadratic expression**. We will investigate quadratic expressions in depth in future lessons.

Glossary

quadratic expression

1. How many small squares are in Step 10?

   (A.) 10

   (B.) 11

   (C.) 90

   (D.) 110

   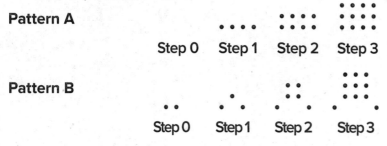

   Step 1   Step 2   Step 3

2. Here are 2 patterns of dots.

   Pattern A

   Step 0   Step 1   Step 2   Step 3

   Pattern B

   Step 0   Step 1   Step 2   Step 3

   a. How many dots will there be in Step 4 of each pattern?

   b. Which pattern shows a quadratic relationship between the step number and the number of dots? Explain how you know.

3. Here are descriptions for how two dot patterns are growing.

   - Pattern A: Step 2 has 10 dots. It grows by 3 dots at each additional step.

   - Pattern B: The total number of dots can be expressed by $2n^2 + 1$, where n is the step number.

   For each pattern, draw a diagram of Step 0 to Step 3.

NAME _____ DATE _____ PERIOD _____

4. Each expression represents the total number of dots in a pattern where n represents the step number.

Select **all** the expressions that represent a quadratic relationship between the step number and the total number of dots. (If you get stuck, consider sketching the first few steps of each pattern as described by the expression.)

(A.) n^2

(B.) $2n$

(C.) $n \cdot n$

(D.) $n + n$

(E.) $n + 2$

(F.) $n \div 2$

5 The function C gives the percentage of homes using only cell phone service x years after 2004. Explain the meaning of each statement. (Lesson 4-3)

a. $C(10) = 35$

b. $C(x) = 10$

c. How is $C(10)$ different from $C(x) = 10$?

6. Here are some lengths, widths, and areas of a garden whose perimeter is 40 feet. (Lesson 6-1)

   a. Complete the table with the missing measurements.

   b. What lengths and widths do you think will produce the largest possible area? Explain how you know.

Length (ft)	Width (ft)	Area (sq ft)
4	16	64
8	12	
10		
12		96
14		
16		64

7. A bacteria population is 10,000 when it is first measured and then doubles each day. (Lesson 5-3)

   a. Use this information to complete the table.

   b. Which is the first day, after the population was originally measured, that the bacteria population is more than 1,000,000?

   c. Write an equation relating p, the bacteria population, to d, the number of days since it was first measured.

d, Time (days)	p, Population (thousands)
0	
1	
2	
5	
10	
d	

8. Graph the solutions to the inequality $7x - 3y \geq 21$. (Lesson 2-21)

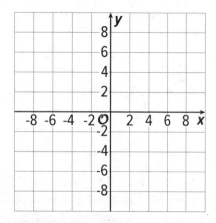

Lesson 6-3

Building Quadratic Functions from Geometric Patterns

NAME _____ DATE _____ PERIOD _____

Learning Goal Let's describe some other geometric patterns.

Warm Up
3.1 Quadratic Expressions and Area

Figure A is a large square. Figure B is a large square with a smaller square removed. Figure C is composed of two large squares with one smaller square added.

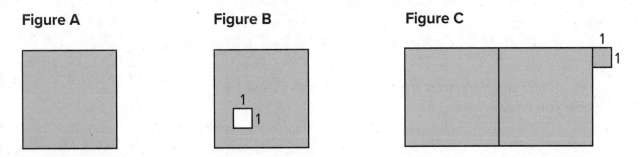

Figure A   **Figure B**   **Figure C**

Write an expression to represent the area of each shaded figure when the side length of the large square is as shown in the first column.

Side Length of Large Square	Area of A	Area of B	Area of C
4			
x			
$4x$			
$(x + 3)$			

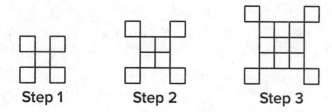

Step 1   **Step 2**   **Step 3**

1. If the pattern continues, what will we see in Step 5 and Step 18?

   a. Sketch or describe the figure in each of these steps.

   b. How many small squares are in each of these steps? Explain how you know.

2. Write an equation to represent the relationship between the step number *n* and the number of squares *y*. Be prepared to explain how each part of your equation relates to the pattern. (If you get stuck, try making a table.)

3. Sketch the first 3 steps of a pattern that can be represented by the equation $y = n^2 - 1$.

NAME _____ DATE _____ PERIOD _____

Are you ready for more?

1. For the original step pattern in the statement, write an equation to represent the relationship between the step number n and the perimeter, P.

2. For the step pattern you created in part 3 of the activity, write an equation to represent the relationship between the step number n and the perimeter, P.

3. Are these linear functions?

Activity

3.3 Growing Steps

1. Sketch the next step in the pattern.

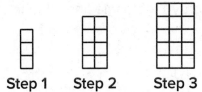

Step 1   Step 2   Step 3

2. Kiran says that the pattern is growing linearly because as the step number goes up by 1, the number of rows and the number of columns also increase by 1. Do you agree? Explain your reasoning.

3. To represent the number of squares after n steps, Diego and Jada wrote different equations. Diego wrote the equation $f(n) = n(n + 2)$. Jada wrote the equation $f(n) = n^2 + 2n$. Are either Diego or Jada correct? Explain your reasoning.

Sometimes a quadratic relationship can be expressed without using a squared term. Let's take this pattern of squares, for example.

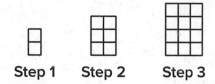

Step 1   Step 2   Step 3

From the first 3 steps, we can see that both the length and the width of the rectangle increase by 1 at each step. Step 1 is a 1-by-2 rectangle, Step 2 is a 2-by-3 rectangle, and Step 3 is a 3-by-4 rectangle. This suggests that Step n is a rectangle with side lengths n and $n + 1$, so the number of squares at Step n is $n(n + 1)$.

This expression may not look like quadratic expressions with a squared term, which we saw in earlier lessons, but if we apply the distributive property, we can see that $n(n + 1)$ is equivalent to $n^2 + n$.

We can also visually show that these expressions are the equivalent by breaking each rectangle into an n-by-n square (the n^2 in the expression) and an n-by-1 rectangle (the n in the expression).

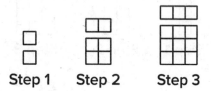

Step 1   Step 2   Step 3

The relationship between the step number and the number of squares can be described by a **quadratic function** f whose input is n and whose output is the number of squares at Step n. We can define f with $f(n) = n(n + 1)$ or $f(n) = n^2 + n$.

Glossary

quadratic function

NAME _____ DATE _____ PERIOD _____

Practice

Building Quadratic Functions from Geometric Patterns

1. Respond to each question.

   a. Sketch or describe the figure in Step 4 and Step 15.

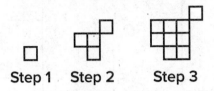

   Step 1 **Step 2** **Step 3**

   b. How many small squares will there be in each of these steps?

   c. Write an equation to represent the relationship between the step number, n, and the number of small squares, y, in each step.

   d. Explain how your equation relates to the pattern.

2. Which expression represents the relationship between the step number n and the total number of small squares in the pattern?

   Step 1 **Step 2** **Step 3**

   A. $n^2 + 1$

   B. $n^2 - 1$

   C. $n^2 - n$

   D. $n^2 + n$

3. Each figure is composed of large squares and small squares. The side length of the large square is x. Write an expression for the area of the shaded part of each figure.

Figure A

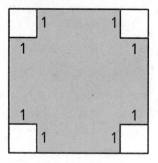

Figure B

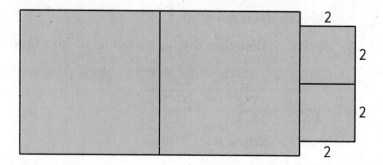

4. Here are a few pairs of positive numbers whose difference is 5. (Lesson 6-1)

   a. Find the product of each pair of numbers. Then, plot some points to show the relationship between the first number and the product.

First Number	Second Number	Product
1	6	
2	7	
3	8	
5	10	
7	12	

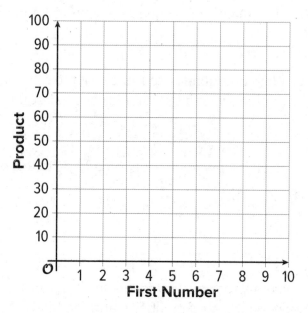

   b. Is the relationship between the first number and the product exponential? Explain how you know.

NAME _____ DATE _____ PERIOD _____

5. Here are some lengths and widths of a rectangle whose perimeter is 20 meters. **(Lesson 6-1)**

   a. Complete the table. What do you notice about the areas?

Length (meters)	Width (meters)	Area (square meters)
1	9	
3	7	
5		
7		
9		

   b. Without calculating, predict whether the area of the rectangle will be greater or less than 25 square meters if the length is 5.25 meters.

   c. On the coordinate plane, plot the points for length and area from your table.

   Do the values change in a linear way? Do they change in an exponential way?

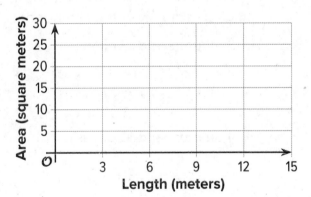

6. Here is a pattern of dots. (Lesson 6-2)

Step 0   Step 1   Step 2   Step 3

Step	Total Number of Dots
0	
1	
2	
3	

a. Complete the table.

b. How many dots will there be in Step 10?

c. How many dots will there be in Step n?

7. Mai has a jar of quarters and dimes. She takes at least 10 coins out of the jar and has less than $2.00. (Lesson 2-25)

   a. Write a system of inequalities that represents the number of quarters, q, and the number of dimes, d, that Mai could have.

   b. Is it possible that Mai has each of the following combinations of coins? If so, explain or show how you know. If not, state which constraint—the amount of money or the number of coins—it does not meet.

      i. 3 quarters and 12 dimes

      ii. 4 quarters and 10 dimes

      iii. 2 quarters and 5 dimes

8. A stadium can seat 63,026 people. For each game, the amount of money that the organization brings in through ticket sales is a function of the number of people, n, in attendance. (Lesson 4-10)

   If each ticket costs $30.00, find the domain and range of this function.

Lesson 6-4

Comparing Quadratic and Exponential Functions

NAME _____ DATE _____ PERIOD _____

Learning Goal Let's compare quadratic and exponential changes and see which one grows faster.

Warm Up
4.1 From Least to Greatest

List these quantities in order, from least to greatest, without evaluating each expression. Be prepared to explain your reasoning.

A. 2^{10} B. 10^2 C. 2^9 D. 9^2

Activity

4.2 Which One Grows Faster?

- In Pattern A, the length and width of the rectangle grow by one small square from each step to the next.

- In Pattern B, the number of small squares doubles from each step to the next.

- In each pattern, the number of small squares is a function of the step number, n.

Pattern A

Step 0   Step 1   Step 2   Step 3

Pattern B

Step 0   Step 1   Step 2   Step 3

1. Write an equation to represent the number of small squares at Step n in Pattern A.

2. Is the function linear, quadratic, or exponential?

1. Write an equation to represent the number of small squares at Step n in Pattern B.

2. Is the function linear, quadratic, or exponential?

NAME _____ DATE _____ PERIOD _____

3. Complete the table:

n, Step Number	f(n), Number of Small Squares
0	
1	
2	
3	
4	
5	
6	
7	
8	

3. Complete the table:

n, Step Number	g(n), Number of Small Squares
0	
1	
2	
3	
4	
5	
6	
7	
8	

How would the two patterns compare if they continue to grow? Make 1–2 observations.

Activity

4.3 Comparing Two More Functions

Here are two functions: $p(x) = 6x^2$ and $q(x) = 3^x$.

Investigate the output of p and q for different values of x. For large enough values of x, one function will have a greater value than the other. Which function will have a greater value as x increases?

Support your answer with tables, graphs, or other representations.

NAME _____ DATE _____ PERIOD _____

Are you ready for more?

1. Jada says that some exponential functions grow more slowly than the quadratic function $f(x) = x^2$ as x increases. Do you agree with Jada? Explain your reasoning.

2. Let $f(x) = x^2$. Could you have an exponential function $g(x) = b^x$ so that $g(x) < f(x)$ for all values of x?

Summary

Comparing Quadratic and Exponential Functions

We have seen that the graphs of quadratic functions can curve upward. Graphs of exponential functions, with base larger than 1, also curve upward. To compare the two, let's look at the quadratic expression $3n^2$ and the exponential expression 2^n.

A table of values shows that $3n^2$ is initially greater than 2^n but 2^n eventually becomes greater.

We also saw an explanation for why exponential growth eventually overtakes quadratic growth.

n	$3n^2$	2^n
1	3	2
2	12	4
3	27	8
4	48	16
5	75	32
6	108	64
7	147	128
8	192	256

- When n increases by 1, the exponential expression 2^n always increases by a factor of 2.

- The quadratic expression $3n^2$ increases by different factors, depending on n, but these factors get smaller. For example, when n increases from 2 to 3, the factor is $\frac{27}{12}$ or 2.25. When n increases from 6 to 7, the factor is $\frac{147}{108}$ or about 1.36. As n increases to larger and larger values, $3n^2$ grows by a factor that gets closer and closer to 1.

A quantity that always doubles will eventually overtake a quantity growing by this smaller factor at each step.

NAME _____ DATE _____ PERIOD _____

Practice
Comparing Quadratic and Exponential Functions

1. The table shows values of the expressions $10x^2$ and 2^x.

   a. Describe how the values of each expression change as x increases.

x	$10x^2$	2^x
1	10	2
2	40	4
3	90	8
4	160	16
8		
10		
12		

   b. Predict which expression will have a greater value when:

   i. x is 8   ii. x is 10   iii. x is 12

   c. Find the value of each expression when x is 8, 10, and 12.

   d. Make an observation about how the values of the two expressions change as x becomes greater and greater.

2. Function f is defined by $f(x) = 1.5^x$. Function g is defined by $g(x) = 500x^2 + 345x$.

   a. Which function is quadratic? Which one is exponential?

   b. The values of which function will eventually be greater for larger and larger values of x?

3. Create a table of values to show that the exponential expression $3(2)^x$ eventually overtakes the quadratic expression $3x^2 + 2x$.

4. The table shows the values of 4^x and $100x^2$ for some values of x.

   Use the patterns in the table to explain why eventually the values of the exponential expression 4^x will overtake the values of the quadratic expression $100x^2$.

x	4^x	$100x^2$
1	4	100
2	16	400
3	64	900
4	256	1600
5	1024	2500

NAME 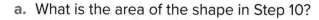 DATE _____ PERIOD _____

5. Here is a pattern of shapes. The area of each small square is 1 sq cm.
   (Lesson 6-2)

   a. What is the area of the shape in Step 10?

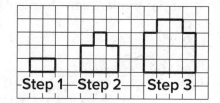

   b. What is the area of the shape in Step *n*?

   c. Explain how you see the pattern growing.

6. A bicycle costs $240 and it loses $\frac{3}{5}$ of its value each year. (Lesson 5-4)

   a. Write expressions for the value of the bicycle, in dollars, after 1, 2, and 3 years.

   b. When will the bike be worth less than $1?

   c. Will the value of the bike ever be 0? Explain your reasoning.

7. A farmer plants wheat and corn. It costs about $150 per acre to plant wheat and about $350 per acre to plant corn. The farmer plans to spend no more than $250,000 planting wheat and corn. The total area of corn and wheat that the farmer plans to plant is less than 1200 acres.

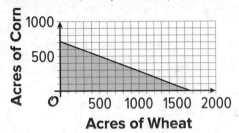

This graph represents the inequality, $150w + 350c \leq 250,000$, which describes the cost constraint in this situation. Let w represent the number of acres of wheat and c represent the number of acres of corn. (Lesson 2-25)

a. The inequality, $w + c < 1,200$ represents the total area constraint in this situation. On the same coordinate plane, graph the solution to this inequality.

b. Use the graphs to find at least two possible combinations of the number of acres of wheat and the number of acres of corn that the farmer could plant.

c. The combination of 400 acres of wheat and 700 acres of corn meets one constraint in the situation but not the other constraint. Which constraint does this meet? Explain your reasoning.

Lesson 6-5

Building Quadratic Functions to Describe Situations (Part 1)

NAME _____ DATE _____ PERIOD _____

Learning Goal Let's measure falling objects.

 Warm Up

5.1 Notice and Wonder: An Interesting Numerical Pattern

Study the table. What do you notice? What do you wonder?

x	0	1	2	3	4	5
y	0	16	64	144	256	400

Activity

5.2 Falling from the Sky

A rock is dropped from the top floor of a 500-foot tall building. A camera captures the distance the rock traveled, in feet, after each second.

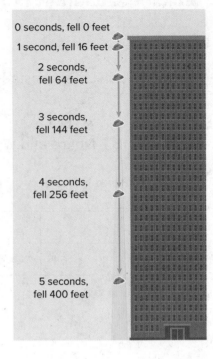

0 seconds, fell 0 feet

1 second, fell 16 feet

2 seconds, fell 64 feet

3 seconds, fell 144 feet

4 seconds, fell 256 feet

5 seconds, fell 400 feet

1. How far will the rock have fallen after 6 seconds? Show your reasoning.

2. Jada noticed that the distances fallen are all multiples of 16. She wrote down:

    $16 = 16 \cdot 1$

    $64 = 16 \cdot 4$

    $144 = 16 \cdot 9$

    $256 = 16 \cdot 16$

    $400 = 16 \cdot 25$

    Then, she noticed that 1, 4, 9, 16, and 25 are 1^2, 2^2, 3^2, 4^2 and 5^2.

    a. Use Jada's observations to predict the distance fallen after 7 seconds. (Assume the building is tall enough that an object dropped from the top of it will continue falling for at least 7 seconds.) Show your reasoning.

    b. Write an equation for the function, with d representing the distance dropped after t seconds.

NAME _____ DATE _____ PERIOD _____

Activity
5.3 Galileo and Gravity

Galileo Galilei, an Italian scientist, and other medieval scholars studied the motion of free-falling objects. The law they discovered can be expressed by the equation $d = 16 \cdot t^2$, which gives the distance fallen in feet, d, as a function of time, t, in seconds.

An object is dropped from a height of 576 feet.

1. How far does it fall in 0.5 seconds?

2. To find out where the object is after the first few seconds after it was dropped, Elena and Diego created different tables.

Elena's table:

Time (seconds)	Distance Fallen (feet)
0	0
1	16
2	64
3	
4	
t	

Diego's table:

Time (seconds)	Distance from The Ground (feet)
0	576
1	560
2	512
3	
4	
t	

a. How are the two tables alike? How are they different?

b. Complete Elena's and Diego's tables. Be prepared to explain your reasoning.

Galileo correctly observed that gravity causes objects to fall in a way where the distance fallen is a quadratic function of the time elapsed. He got a little carried away, however, and assumed that a hanging rope or chain could also be modeled by a quadratic function.

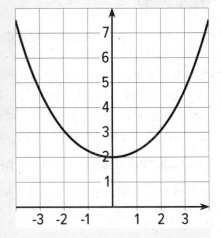

Here is a graph of such a shape (called a catenary) along with a table of approximate values.

x	-4	-3	-2	-1	0	1	2	3	4
y	7.52	4.70	3.09	2.26	2	2.26	3.09	4.70	7.52

Show that an equation of the form $y = ax^2 + b$ cannot model this data well.

NAME _____ DATE _____ PERIOD _____

Summary
Building Quadratic Functions to Describe Situations (Part 1)

The distance traveled by a falling object in a given amount of time is an example of a quadratic function. Galileo is said to have dropped balls of different mass from the Leaning Tower of Pisa, which is about 190 feet tall, to show that they travel the same distance in the same time. In fact the equation $d = 16t^2$ models the distance d, in feet, that the cannonball falls after t seconds, no matter what its mass.

Because $16 \cdot 4^2 = 256$, and the tower is only 190 feet tall, the cannonball hits the ground before 4 seconds.

Here is a table showing how far the cannonball has fallen over the first few seconds.

Here are the time and distance pairs plotted on a coordinate plane:

Time (seconds)	Distance Fallen (feet)
0	0
1	16
2	64
3	144

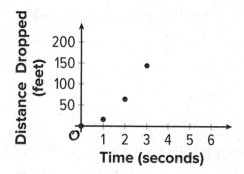

Notice that the distance fallen is increasing each second. The average rate of change is increasing each second, which means that the cannonball is speeding up over time. This comes from the influence of gravity, which is represented by the quadratic expression $16t^2$. It is the exponent 2 in that expression that makes it increase by larger and larger amounts.

Another way to study the change in the position of the cannonball is to look at its distance from the ground as a function of time.

Here is a table showing the distance from the ground in feet at 0, 1, 2, and 3 seconds.

Here are the time and distance pairs plotted on a graph:

Time (seconds)	Distance from the Ground (feet)
0	190
1	174
2	126
3	46

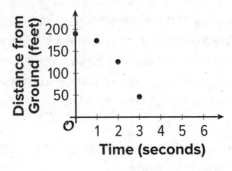

The expression that defines the distance from the ground as a function of time is $190 - 16t^2$. It tells us that the cannonball's distance from the ground is 190 feet before it is dropped and has decreased by $16t^2$ when t seconds have passed.

NAME _____ DATE _____ PERIOD _____

Practice
Building Quadratic Functions to Describe Situations (Part 1)

1. A rocket is launched in the air and its height, in feet, is modeled by the function h. Here is a graph representing h.

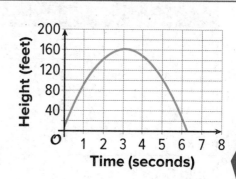

   Select **all** true statements about the situation.

   (A.) The rocket is launched from a height less than 20 feet above the ground.

   (B.) The rocket is launched from about 20 feet above the ground.

   (C.) The rocket reaches its maximum height after about 3 seconds.

   (D.) The rocket reaches its maximum height after about 160 seconds.

   (E.) The maximum height of the rocket is about 160 feet.

2. A baseball travels d meters t seconds after being dropped from the top of a building. The distance traveled by the baseball can be modeled by the equation $d = 5t^2$.

   a. Complete the table and plot the data on the coordinate plane.

t (seconds)	d (meters)
0	
0.5	
1	
1.5	
2	

   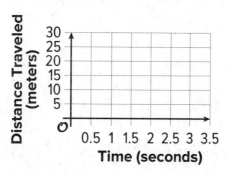

   b. Is the baseball traveling at a constant speed? Explain how you know.

3. A rock is dropped from a bridge over a river. Which table could represent the distance in feet fallen as a function of time in seconds?

Table A

Time (seconds)	Distance Fallen (feet)
0	0
1	48
2	96
3	144

Table B

Time (seconds)	Distance Fallen (feet)
0	0
1	16
2	64
3	144

Table C

Time (seconds)	Distance Fallen (feet)
0	180
1	132
2	84
3	36

Table D

Time (seconds)	Distance Fallen (feet)
0	180
1	164
2	116
3	36

A. Table A

B. Table B

C. Table C

D. Table D

NAME _____ DATE _____ PERIOD _____

4. Determine whether $5n^2$ or 3^n will have the greater value when: **(Lesson 6-4)**

   a. $n = 1$

   b. $n = 3$

   c. $n = 5$

5. Select **all** of the expressions that give the number of small squares in Step n. **(Lesson 6-3)**

   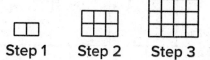

   Step 1   Step 2   Step 3

   (A.) $2n$

   (B.) n^2

   (C.) $n + 1$

   (D.) $n^2 + 1$

   (E.) $n(n + 1)$

   (F.) $n^2 + n$

   (G.) $n + n + 1$

6. A small ball is dropped from a tall building. Which equation could represent the ball's height, h, in feet, relative to the ground, as a function of time, t, in seconds?

   (A.) $h = 100 - 16t$

   (B.) $h = 100 - 16t^2$

   (C.) $h = 100 - 16^t$

   (D.) $h = 100 - \dfrac{16}{t}$

7. Use the rule for function f to draw its graph. (Lesson 4-12)

$$f(x) = \begin{cases} 2, & -5 \le x < -2 \\ 6, & -2 \le x < 4 \\ x, & 4 \le x < 8 \end{cases}$$

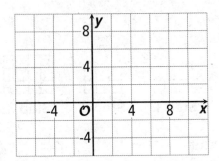

8. Diego claimed that $10 + x^2$ is always greater than 2^x and used this table as evidence. (Lesson 6-4)

Do you agree with Diego?

x	$10 + x^2$	2^x
1	11	2
2	14	4
3	19	8
4	26	16

9. The table shows the height, in centimeters, of the water in a swimming pool at different times since the pool started to be filled. (Lesson 5-2)

a. Does the height of the water increase by the same amount each minute? Explain how you know.

b. Does the height of the water increase by the same factor each minute? Explain how you know.

Minutes	Height
0	150
1	150.5
2	151
3	151.5

Lesson 6-6

Building Quadratic Functions to Describe Situations (Part 2)

NAME _____ DATE _____ PERIOD _____

Learning Goal Let's look at the objects being launched in the air.

Warm Up
6.1 Sky Bound

A cannon is 10 feet off the ground. It launches a cannonball straight up with a velocity of 406 feet per second.

Imagine that there is no gravity and that the cannonball continues to travel upward with the same velocity.

1. Complete the table with the heights of the cannonball at different times.

Seconds	0	1	2	3	4	5	t
Distance Above Ground (feet)	10						

2. Write an equation to model the distance in feet, d, of the ball t seconds after it was fired from the cannon if there was no gravity.

Activity

6.2 Tracking a Cannonball

Earlier, you completed a table that represents the height of a cannonball, in feet, as a function of time, in seconds, if there was no gravity.

1. This table shows the actual heights of the ball at different times.

Seconds	0	1	2	3	4	5
Distance Above Ground (Feet)	10	400	758	1,084	1,378	1,640

Compare the values in this table with those in the table you completed earlier. Make at least 2 observations.

2. Respond to each question.

a. Plot the two sets of data you have on the same coordinate plane.

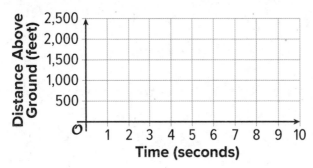

NAME _____ DATE _____ PERIOD _____

b. How are the two graphs alike? How are they different?

3. Write an equation to model the actual distance *d*, in feet, of the ball *t* seconds after it was fired from the cannon. If you get stuck, consider the differences in distances and the effects of gravity from a previous lesson.

Activity

6.3 Graphing Another Cannonball

The function defined by $d = 50 + 312t - 16t^2$ gives the height in feet of a cannonball *t* seconds after the ball leaves the cannon.

1. What do the terms 50, 312*t*, and -16t^2 tell us about the cannonball?

2. Use graphing technology to graph the function. Adjust the graphing window to the following boundaries: $0 < x < 25$ and $0 < y < 2{,}000$.

3. Observe the graph and:

   a. Describe the shape of the graph. What does it tell us about the movement of the cannonball?

   b. Estimate the maximum height the ball reaches. When does this happen?

   c. Estimate when the ball hits the ground.

4. What domain is appropriate for this function? Explain your reasoning.

Are you ready for more?

If the cannonball were fired at 800 feet per second, would it reach a mile in height? Explain your reasoning.

NAME _____ DATE _____ PERIOD _____

Summary

Building Quadratic Functions to Describe Situations (Part 2)

In this lesson, we looked at the height of objects that are launched upward and then come back down because of gravity.

An object is thrown upward from a height of 5 feet with a velocity of 60 feet per second. Its height $h(t)$ in feet after t seconds is modeled by the function $h(t) = 5 + 60t - 16t^2$.

- The linear expression $5 + 60t$ represents the height the object would have at time t if there were no gravity. The object would keep going up at the same speed at which it was thrown. The graph would be a line with a slope of 60 which relates to the constant speed of 60 feet per second.

- The expression $-16t^2$ represents the effect of gravity, which eventually causes the object to slow down, stop, and start falling back again.

Notice the graph intersects the vertical axis at 5, which means the object was thrown into the air from 5 feet off the ground. The graph indicates that the object reaches its peak height of about 60 feet after a little less than 2 seconds. That peak is the point on the graph where the function reaches a maximum value. At that point, the curve changes direction, and the output of the function changes from increasing to decreasing. We call that point the **vertex** of the graph.

Here is the graph of h.

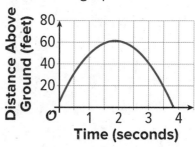

The graph representing any quadratic function is a special kind of "U" shape called a *parabola*. You will learn more about the geometry of parabolas in a future course. Every parabola has a vertex, because there is a point where it changes direction—from increasing to decreasing, or the other way around.

The object hits the ground a little before 4 seconds. That time corresponds to the horizontal intercept of the graph. An input value that produces an output of 0 is called a **zero** of the function. A zero of the function h is approximately 3.8, because $h(3.8) \approx 0$.

In this situation, input values less than 0 seconds or more than about 3.8 seconds would not be meaningful, so an appropriate domain for this function would include all values of t between 0 and about 3.8.

Glossary

vertex (of a graph)
zero (of a function)

Practice
Building Quadratic Functions to Describe Situations (Part 2)

1. The height of a diver above the water, is given by $h(t) = -5t^2 + 10t + 3$, where t is time measured in seconds and $h(t)$ is measured in meters. Select **all** statements that are true about the situation.

   (A.) The diver begins 5 meters above the water.

   (B.) The diver begins 3 meters above the water.

   (C.) The function has 1 zero that makes sense in this situation.

   (D.) The function has 2 zeros that make sense in this situation.

   (E.) The graph that represents h starts at the origin and curves upward.

   (F.) The diver begins at the same height as the water level.

2. The height of a baseball, in feet, is modeled by the function h given by the equation $h(t) = 3 + 60t - 16t^2$. The graph of the function is shown.

   a. About when does the baseball reach its maximum height?

   b. About how high is the maximum height of the baseball?

   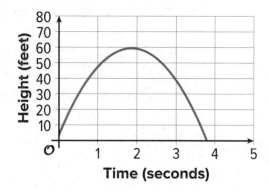

   c. About when does the ball hit the ground?

3. *Technology required.* Two rocks are launched straight up in the air. The height of Rock A is given by the function f, where $f(t) = 4 + 30t - 16t^2$. The height of Rock B is given by g, where $g(t) = 5 + 20t - 16t^2$. In both functions, t is time measured in seconds and height is measured in feet.

   Use graphing technology to graph both equations. Determine which rock hits the ground first and explain how you know.

4. Each expression represents an object's distance from the ground in meters as a function of time, t, in seconds.

   Object A: $-5t^2 + 25t + 50$

   Object B: $-5t^2 + 50t + 25$

   a. Which object was launched with the greatest vertical speed?

   b. Which object was launched from the greatest height?

5. Tyler is building a pen for his rabbit on the side of the garage. He needs to fence in three sides and wants to use 24 ft of fencing. **(Lesson 6-1)**

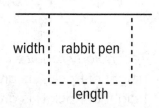

    a. The table shows some possible lengths and widths. Complete each area.

    b. Which length and width combination should Tyler choose to give his rabbit the most room?

Length (ft)	Width (ft)	Area (sq ft)
8	8	
10	7	
12	6	
14	5	
16	4	

6. Here is a pattern of dots. **(Lesson 6-2)**

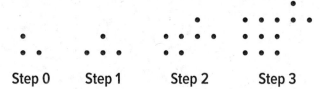

Step 0   Step 1   Step 2   Step 3

Step	Total Number of Dots
0	
1	
2	
3	

    a. Complete the table.

    b. How many dots will there be in Step 10?

    c. How many dots will there be in Step n?

NAME _____ DATE _____ PERIOD _____

7. The function f is defined by $f(x) = 2^x$ and the function g is defined by $g(x) = x^2 + 16$. **(Lesson 6-4)**

   a. Find the values of f and g when x is 4, 5, and 6.

   b. Are the values of $f(x)$ always greater than the values of $g(x)$ for all x? Explain how you know.

8. Han accidentally drops his water bottle from the balcony of his apartment building. The equation $d = 32 - 5t^2$ gives the distance from the ground, d, in meters after t seconds. **(Lesson 6-5)**

   a. Complete the table and plot the data on the coordinate plane.

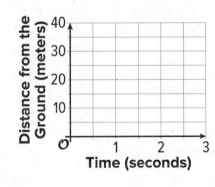

t (seconds)	d (meters)
0	
0.5	
1	
1.5	
2	

   b. Is the water bottle falling at a constant speed? Explain how you know.

9. The graph shows how much insulin, in micrograms (mcg), is in a patient's body after receiving an injection. **(Lesson 5-6)**

a. Write an equation giving the number of mcg of insulin, m, in the patient's body h hours after receiving the injection. Assume the amount of insulin continues to decay exponentially.

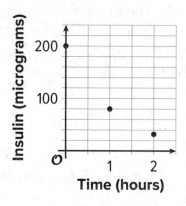

b. After 3 hours, will the patient still have at least 10 mcg of insulin in their body? Explain how you know.

Lesson 6-7

Building Quadratic Functions to Describe Situations (Part 3)

NAME _____ DATE _____ PERIOD _____

Learning Goal Let's look at how to maximize revenue.

Warm Up

7.1 Which One Doesn't Belong: Graphs of Four Functions

Which one doesn't belong?

A

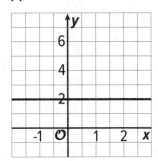

B

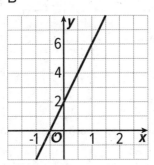

C

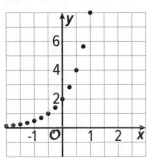

D

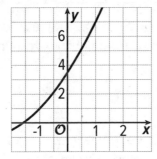

A company that sells movies online is deciding how much to charge customers to download a new movie. Based on data from previous sales, the company predicts that if they charge x dollars for each download, then the number of downloads, in thousands, is $18 - x$.

1. Complete the table to show the predicted number of downloads at each listed price. Then, find the revenue at each price. The first row has been completed for you.

Price (dollars per download)	Number of Downloads (thousands)	Revenue (thousands of dollars)
3	15	45
5		
10		
12		
15		
18		
x		

2. Is the relationship between the price of the movie and the revenue (in thousands of dollars) quadratic? Explain how you know.

3. Plot the points that represent the revenue, r, as a function of the price of one download in dollars, x.

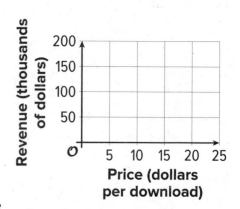

4. What price would you recommend the company charge for a new movie? Explain your reasoning.

NAME _____ DATE _____ PERIOD _____

Are you ready for more?

The function that uses the price (in dollars per download) x to determine the number of downloads (in thousands) $18 - x$ is an example of a demand function and its graph is known. Economists are interested in factors that can affect the demand function and therefore the price suppliers wish to set.

1. What are some things that could increase the number of downloads predicted for the same given prices?

2. If the demand shifted so that we predicted $20 - x$ thousand downloads at a price of x dollars per download, what do you think will happen to the price that gives the maximum revenue? Check what actually happens.

Activity
7.3 Domain, Vertex, and Zeros

Here are 4 sets of descriptions and equations that represent some familiar quadratic functions. The graphs show what a graphing technology may produce when the equations are graphed. For each function:

- Describe a domain that is appropriate for the situation. Think about any upper or lower limits for the input, as well as whether all numbers make sense as the input. Then, describe how the graph should be modified to show the domain that makes sense.

- Identify or estimate the vertex on the graph. Describe what it means in the situation.

- Identify or estimate the zeros of the function. Describe what it means in the situation.

1. The area of a rectangle with a perimeter of 25 meters and a side length x: $A(x) = x \cdot \dfrac{(25 - 2x)}{2}$

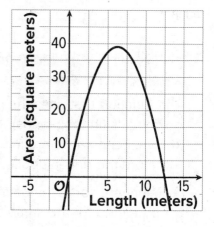

   - Domain:

   - Vertex:

   - Zeros:

2. The number of squares as a function of step number n: $f(n) = n^2 + 4$

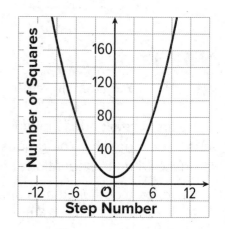

   - Domain:

   - Vertex:

   - Zeros

NAME _____ DATE _____ PERIOD _____

3. The distance in feet that an object has fallen t seconds after being dropped: $g(t) = 16t^2$

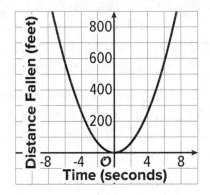

• Domain:

• Vertex:

• Zeros:

4. The height in feet of an object t seconds after being dropped: $h(t) = 576 - 16t^2$

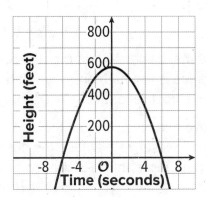

• Domain:

• Vertex:

• Zeros:

Summary
Building Quadratic Functions to Describe Situations (Part 3)

Quadratic functions often come up when studying revenue. (*Revenue* means the money collected when someone sells something.)

Suppose we are selling raffle tickets and deciding how much to charge for each ticket. When the price of the tickets is higher, typically fewer tickets will be sold.

Let's say that with a price of d dollars, it is possible to sell $600 - 75d$ tickets. We can find the revenue by multiplying the price by the number of tickets expected to be sold. A function that models the revenue r collected is $r(d) = d(600 - 75d)$. Here is a graph that represents the function.

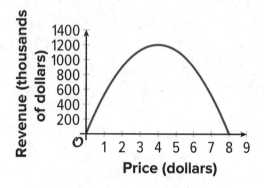

It makes sense that the revenue goes down after a certain point, since if the price is too high nobody will buy a ticket. From the graph, we can tell that the greatest revenue, $1,200, comes from selling the tickets for $4 each.

We can also see that the domain of the function r is between 0 and 8. This makes sense because the cost of the tickets can't be negative, and if the price were more than $8, the model does not work, as the revenue collected cannot be negative. (A negative revenue would mean the number of tickets sold is negative, which is not possible.)

NAME _____ DATE _____ PERIOD _____

Practice
Building Quadratic Functions to Describe Situations (Part 3)

1. Based on past musical productions, a theater predicts selling $400 - 8p$ tickets when each ticket is sold at p dollars.

   a. Complete the table to find out how many tickets the theater expects to sell and what revenues it expects to receive at the given ticket prices.

Ticket Price (dollars)	Number of Tickets Sold	Revenue (dollars)
5		
10		
15		
20		
30		
45		
50		
p		

   b. For which ticket prices will the theater earn no revenue? Explain how you know.

   c. At what ticket prices should the theater sell the tickets if it must earn at least $3,200 in revenue to break even (to not lose money) on the musical production? Explain how you know.

2. A company sells running shoes. If the price of a pair of shoes in dollars is p, the company estimates that it will sell $50,000 - 400p$ pairs of shoes.

   Write an expression that represents the revenue in dollars from selling running shoes if a pair of shoes is priced at p dollars.

3. The function *f* represents the revenue in dollars the school can expect to receive if it sells $220 - 12x$ coffee mugs for *x* dollars each.

   Here is the graph of *f*.

   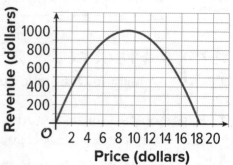

   Select **all** the statements that describe this situation.

   (A.) At $2 per coffee mug, the revenue will be $196.

   (B.) The school expects to sell 160 mugs if the price is $5 each.

   (C.) The school will lose money if it sells the mugs for more than $10 each.

   (D.) The school will earn about $1,000 if it sells the mugs for $10 each.

   (E.) The revenue will be more than $700 if the price is between $4 and $14.

   (F.) The expected revenue will increase if the price per mug is greater than $10.

4. Respond to each question.

   a. Write an equation to represent the relationship between the step number, *n*, and the number of small squares, *y*. (Lesson 6-3)

   Briefly describe how each part of the equation relates to the pattern.

   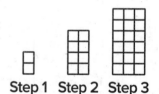

   Step 1   Step 2   Step 3

   b. Is the relationship between the step number and number of small squares quadratic? Explain how you know.

NAME _____ DATE _____ PERIOD _____

5. *Technology required.* A small marshmallow is launched straight up in the air with a slingshot. The function h, given by the equation $h(t) = 5 + 20t - 5t^2$, describes the height of the marshmallow in meters as a function of time, t, in seconds since it was launched. **(Lesson 6-5)**

   a. Use graphing technology to graph the function h.

   b. About when does the marshmallow reach its maximum height?

   c. About how long does it take before the marshmallow hits the ground?

   d. What domain makes sense for the function h in this situation?

6. A rock is dropped from a bridge over a river. Which graph could represent the distance fallen, in feet, as a function of time in seconds? **(Lesson 6-5)**

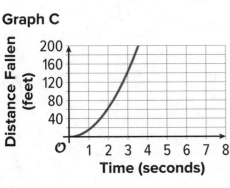

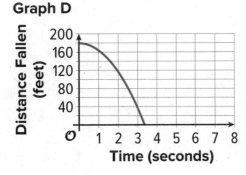

(A.) Graph A

(B.) Graph B

(C.) Graph C

(D.) Graph D

7. A bacteria population, p, is modeled by the equation $p = 100{,}000 \cdot 2^d$, where d is the number of days since the population was first measured. (Lesson 5-7)

Select **all** statements that are true in this situation.

(A.) $100{,}000 \cdot 2^{-2}$ represents the bacteria population 2 days before it was first measured.

(B.) The bacteria population 3 days before it was first measured was 800,000.

(C.) The population was more than 1,000 one week before it was first measured.

(D.) The population was more than 1,000,000 one week after it was first measured.

(E.) The bacteria population 4 days before it was first measured was 6,250.

Lesson 6-8

Equivalent Quadratic Expressions

NAME _____ DATE _____ PERIOD _____

Learning Goal Let's use diagrams to help us rewrite quadratic expressions.

 Warm Up
8.1 Diagrams of Products

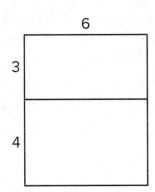

1. Explain why the diagram shows that
$6(3 + 4) = 6 \cdot 3 + 6 \cdot 4$.

2. Draw a diagram to show that
$5(x + 2) = 5x + 10$.

 Activity
8.2 Drawing Diagrams to Represent More Products

Applying the distributive property to multiply out the factors of, or expand, $4(x + 2)$ gives us $4x + 8$, so we know the two expressions are equivalent. We can use a rectangle with side lengths $(x + 2)$ and 4 to illustrate the multiplication.

	x	2
4	$4x$	8

1. Draw a diagram to show that $n(2n + 5)$ and $2n^2 + 5n$ are equivalent expressions.

2. For each expression, use the distributive property to write an equivalent expression. If you get stuck, consider drawing a diagram.

   a. $6\left(\frac{1}{3}n + 2\right)$

   b. $p(4p + 9)$

   c. $5r\left(r + \frac{3}{5}\right)$

   d. $(0.5w + 7)w$

Activity

8.3 Using Diagrams to Find Equivalent Quadratic Expressions

1. Here is a diagram of a rectangle with side lengths $x + 1$ and $x + 3$. Use this diagram to show that $(x + 1)(x + 3)$ and $x^2 + 4x + 3$ are equivalent expressions.

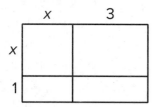

2. Draw diagrams to help you write an equivalent expression for each of the following:

   a. $(x + 5)^2$

   b. $2x(x + 4)$

   c. $(2x + 1)(x + 3)$

   d. $(x + m)(x + n)$

3. Write an equivalent expression for each expression without drawing a diagram:

   a. $(x + 2)(x + 6)$

   b. $(x + 5)(2x + 10)$

NAME _____ DATE _____ PERIOD _____

Are you ready for more?

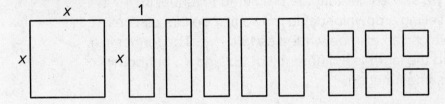

1. Is it possible to arrange an x by x square, five x by 1 rectangles and six 1 by 1 squares into a single large rectangle? Explain or show your reasoning.

2. What does this tell you about an equivalent expression for $x^2 + 5x + 6$?

3. Is there a different non-zero number of 1 by 1 squares that we could have used instead that would allow us to arrange the combined figures into a single large rectangle?

A quadratic function can often be defined by many different but equivalent expressions. For example, we saw earlier that the predicted revenue, in thousands of dollars, from selling a downloadable movie at x dollars can be expressed with $x(18 - x)$, which can also be written as $18x - x^2$. The former is a product of x and $18 - x$, and the latter is a difference of $18x$ and x^2, but both expressions represent the same function.

Sometimes a quadratic expression is a product of two factors that are each a linear expression, for example $(x + 2)(x + 3)$. We can write an equivalent expression by thinking about each factor, the $(x + 2)$ and $(x + 3)$, as the side lengths of a rectangle, and each side length decomposed into a variable expression and a number.

	x	2
x	x^2	$2x$
3	$3x$	6

Multiplying $(x + 2)$ and $(x + 3)$ gives the area of the rectangle. Adding the areas of the four sub-rectangles also gives the area of the rectangle. This means that $(x + 2)(x + 3)$ is equivalent to $x^2 + 2x + 3x + 6$, or to $x^2 + 5x + 6$.

Notice that the diagram illustrates the distributive property being applied. Each term of one factor (say, the x and the 2 in $x + 2$) is multiplied by every term in the other factor (the x and the 3 in $x + 3$).

In general, when a quadratic expression is written in the form of $(x + p)(x + q)$, we can apply the distributive property to rewrite it as $x^2 + px + qx + pq$ or $x^2 + (p + q)x + pq$.

$$(x + 2)(x + 3)$$

$$= x(x + 3) + 2(x + 3)$$

$$= x^2 + 3x + 2x + (2)(3)$$

$$= x^2 + (3 + 2)x + (2)(3)$$

NAME _____ DATE _____ PERIOD _____

Practice
Equivalent Quadratic Expressions

1. Draw a diagram to show that $(2x + 5)(x + 3)$ is equivalent to $2x^2 + 11x + 15$.

2. Match each quadratic expression that is written as a product with an equivalent expression that is expanded.

   A. $(x + 2)(x + 6)$ 1. $x^2 + 12x + 32$

   B. $(2x + 8)(x + 2)$ 2. $2x^2 + 10x + 12$

   C. $(x + 8)(x + 4)$ 3. $2x^2 + 12x + 16$

   D. $(x + 2)(2x + 6)$ 4. $x^2 + 8x + 12$

3. Select **all** expressions that are equivalent to $x^2 + 4x$.

   (A.) $x(x + 4)$ (C.) $(x + x)(x + 4)$ (E.) $(x + 4)x$

   (B.) $(x + 2)^2$ (D.) $(x + 2)^2 - 4$

4. Tyler drew a diagram to expand $(x + 5)(2x + 3)$.

   a. Explain Tyler's mistake.

	$2x$	3
x	$2x^2$	$3x$
5	$7x$	8

   b. What is the correct expanded form of $(x + 5)(2x + 3)$?

5. Explain why the values of the exponential expression 3^x will eventually overtake the values of the quadratic expression $10x^2$. **(Lesson 6-4)**

6. A baseball travels d meters t seconds after being dropped from the top of a building. The distance traveled by the baseball can be modeled by the equation $d = 5t^2$. **(Lesson 6-5)**

Which graph could represent this situation? Explain how you know.

Graph A

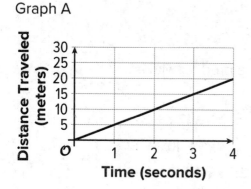

Graph B

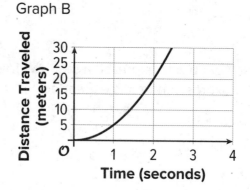

7. Consider a function q defined by $q(x) = x^2$. Explain why negative values are not included in the range of q. **(Lesson 4-10)**

NAME _____ DATE _____ PERIOD _____

8. Based on past concerts, a band predicts selling $600 - 10p$ concert tickets when each ticket is sold at p dollars. **(Lesson 6-7)**

   a. Complete the table to find out how many concert tickets the band expects to sell and what revenues it expects to receive at the given ticket prices.

Ticket Price (dollars)	Number of Tickets	Revenue (dollars)
10		
15		
20		
30		
35		
45		
50		
60		
p		

   b. In this model, at what ticket prices will the band earn no revenue at all?

   c. At what ticket prices should the band sell the tickets if it must earn at least 8,000 dollars in revenue to break even (to not lose money) on a given concert. Explain how you know.

9. A population of bears decreases exponentially. The population was first measured in 2010. **(Lesson 5-8)**

   a. What is the annual factor of decrease for the bear population? Explain how you know.

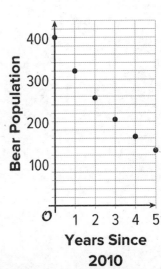

Bear Population vs. Years Since 2010

   b. Using function notation, represent the relationship between the bear population, b, and the number of years since the population was first measured, t. That is, find a function, f, so that $b = f(t)$.

10. Equations defining functions a, b, c, d, and f are shown here.

Select **all** the equations that represent exponential functions. (Lesson 5-8)

(A.) $a(x) = 2^3 \cdot x$

(B.) $b(t) = \left(\frac{2}{3}\right)^t$

(C.) $c(m) = \frac{1}{5} \cdot 2^m$

(D.) $d(x) = 3x^2$

(E.) $f(t) = 3 \cdot 2^t$

Lesson 6-9

Standard Form and Factored Form

NAME _____ DATE _____ PERIOD _____

Learning Goal Let's write quadratic expressions in different forms.

 ## Warm Up
9.1 Math Talk: Opposites Attract

Solve each equation mentally.

$40 - 8 = 40 + n$

$25 + \text{-}100 = 25 - n$

$3 - \dfrac{1}{2} = 3 + n$

$72 - n = 72 + 6$

 ## Activity
9.2 Finding Products of Differences

1. Show that $(x - 1)(x - 1)$ and $x^2 - 2x + 1$ are equivalent expressions by drawing a diagram or applying the distributive property. Show your reasoning.

2. For each expression, write an equivalent expression. Show your reasoning.

a. $(x + 1)(x - 1)$

b. $(x - 2)(x + 3)$

c. $(x - 2)^2$

 ## Activity

9.3 What Is the Standard Form? What Is the Factored Form?

The quadratic expression $x^2 + 4x + 3$ is written in **standard form**.

Here are some other quadratic expressions. The expressions on the left are written in standard form and the expressions on the right are not.

Written in standard form:	Not written in standard form:
$x^2 - 1$	$(2x + 3)x$
$x^2 + 9x$	$(x + 1)(x - 1)$
$\frac{1}{2}x^2$	$3(x - 2)^2 + 1$
$4x^2 - 2x + 5$	$-4(x^2 + x) + 7$
$-3x^2 - x + 6$	$(x + 8)(-x + 5)$
$1 - x^2$	

NAME _____ DATE _____ PERIOD _____

1. What are some characteristics of expressions in standard form?

2. $(x + 1)(x - 1)$ and $(2x + 3)x$ in the right column are quadratic expressions written in **factored form**. Why do you think that form is called factored form?

Are you ready for more?

Which quadratic expression can be described as being both standard form and factored form? Explain how you know.

Summary
Standard Form and Factored Form

A quadratic function can often be represented by many equivalent expressions. For example, a quadratic function f might be defined by $f(x) = x^2 + 3x + 2$. The quadratic expression $x^2 + 3x + 2$ is called the **standard form**, the sum of a multiple of x^2 and a linear expression ($3x + 2$ in this case).

In general, standard form is

$$ax^2 + bx + c$$

We refer to a as the coefficient of the squared term x^2, b as the coefficient of the linear term x, and c as the constant term.

The function f can also be defined by the equivalent expression $(x + 2)(x + 1)$. When the quadratic expression is a product of two factors where each one is a linear expression, this is called the **factored form**.

An expression in factored form can be rewritten in standard form by expanding it, which means multiplying out the factors. In a previous lesson we saw how to use a diagram and to apply the distributive property to multiply two linear expressions, such as $(x + 3)(x + 2)$. We can do the same to expand an expression with a sum and a difference, such as $(x + 5)(x - 2)$, or to expand an expression with two differences, for example, $(x - 4)(x - 1)$.

To represent $(x - 4)(x - 1)$ with a diagram, we can think of subtraction as adding the opposite:

	x	-4
x	x^2	-4x
-1	-x	-4x

$$(x - 4)(x - 1)$$
$$= (x + -4)(x + -1)$$
$$= x(x + -1) + -4(x + -1)$$
$$= x^2 + -1x + -4x + (-4)(-1)$$
$$= x^2 + -5x + 4$$
$$= x^2 - 5x + 4$$

Glossary

factored form (of a quadratic expression)
standard form (of a quadratic expression)

NAME _____ DATE _____ PERIOD _____

Practice
Standard Form and Factored Form

1. Write each quadratic expression in standard form. Draw a diagram if needed.

   a. $(x + 4)(x - 1)$

   b. $(2x - 1)(3x - 1)$

2. Consider the expression $8 - 6x + x^2$.

   a. Is the expression in standard form? Explain how you know.

   b. Is the expression equivalent to $(x - 4)(x - 2)$? Explain how you know.

3. Which quadratic expression is written in standard form?

   Ⓐ $(x + 3)x$

   Ⓑ $(x + 4)^2$

   Ⓒ $-x^2 - 5x + 7$

   Ⓓ $x^2 + 2(x + 3)$

4. Explain why $3x^2$ can be said to be in both standard form and factored form.

5. Jada dropped her sunglasses from a bridge over a river. Which equation could represent the distance y fallen in feet as a function of time, t, in seconds? (Lesson 6-5)

(A.) $y = 16t^2$

(B.) $y = 48t$

(C.) $y = 180 - 16t^2$

(D.) $y = 180 - 48t$

6. A football player throws a football. The function h given by $h(t) = 6 + 75t - 16t^2$ describes the football's height in feet t seconds after it is thrown. (Lesson 6-6)

Select **all** the statements that are true about this situation.

(A.) The football is thrown from ground level.

(B.) The football is thrown from 6 feet off the ground.

(C.) In the function, $-16t^2$ represents the effect of gravity.

(D.) The outputs of h decrease then increase in value.

(E.) The function h has 2 zeros that make sense in this situation.

(F.) The vertex of the graph of h gives the maximum height of the football.

7. *Technology required.* Two rocks are launched straight up in the air.

- The height of Rock A is given by the function f, where $f(t) = 4 + 30t - 16t^2$.

- The height of Rock B is given by function g, where $g(t) = 5 + 20t - 16t^2$.

In both functions, t is time measured in seconds and height is measured in feet. Use graphing technology to graph both equations. (Lesson 6-6)

a. What is the maximum height of each rock?

b. Which rock reaches its maximum height first? Explain how you know.

NAME _____ DATE _____ PERIOD _____

8. The graph shows the number of grams of a radioactive substance in a sample at different times after the sample was first analyzed. (Lesson 5-10)

a. What is the average rate of change for the substance during the 10 year period?

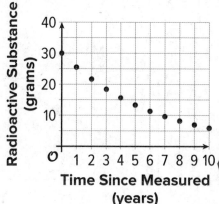

b. Is the average rate of change a good measure for the change in the radioactive substance during these 10 years? Explain how you know.

9. Each day after an outbreak of a new strain of the flu virus, a public health scientist receives a report of the number of new cases of the flu reported by area hospitals.

Time Since Outbreak in Days	2	3	4	5	6	7
Number of New Cases of the Flu	20	28	38	54	75	105

Would a linear or exponential model be more appropriate for this data? Explain how you know. (Lesson 5-11)

10. $A(t)$ is a model for the temperature in Aspen, Colorado, t months after the start of the year. $M(t)$ is a model for the temperature in Minneapolis, Minnesota, t months after the start of the year. Temperature is measured in degrees Fahrenheit. (Lesson 4-9)

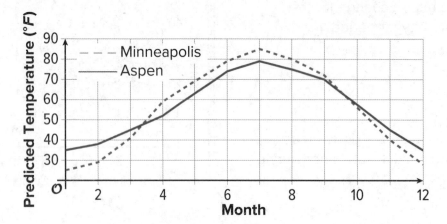

a. What does $A(8)$ mean in this situation? Estimate $A(8)$.

b. Which city has a higher predicted temperature in February?

c. Are the two cities' predicted temperatures ever the same? If so, when?

Lesson 6-10

Graphs of Functions in Standard and Factored Forms

NAME _____ DATE _____ PERIOD _____

Learning Goal Let's find out what quadratic expressions in standard and factored forms can reveal about the properties of their graphs.

 ## Warm Up
10.1 A Linear Equation and Its Graph

Here is a graph of the equation $y = 8 - 2x$.

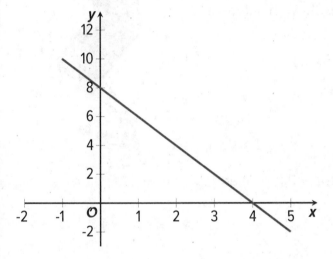

1. Where do you see the 8 from the equation in the graph?

2. Where do you see the -2 from the equation in the graph?

3. What is the x-intercept of the graph? How does this relate to the equation?

 Activity

10.2 Revisiting Projectile Motion

In an earlier lesson, we saw that an equation such as $h(t) = 10 + 78t - 16t^2$ can model the height of an object thrown upward from a height of 10 feet with a vertical velocity of 78 feet per second.

1. Is the expression $10 + 78t - 16t^2$ written in standard form? Explain how you know.

2. Jada said that the equation $g(t) = (\text{-}16t - 2)(t - 5)$ also defines the same function, written in factored form. Show that Jada is correct.

3. Here is a graph representing both $g(t) = (\text{-}16t - 2)(t - 5)$ and $h(t) = 10 + 78t - 16t^2$.

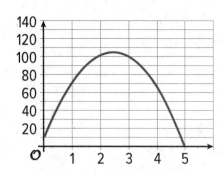

   a. Identify or approximate the vertical and horizontal intercepts.

   b. What do each of these points mean in this situation?

NAME _____ DATE _____ PERIOD _____

Activity

10.3 Relating Expressions and Their Graphs

Here are pairs of expressions in standard and factored forms. Each pair of expressions define the same quadratic function, which can be represented with the given graph.

1. Identify the x-intercepts and the y-intercept of each graph.

Function f

$x^2 + 4x + 3$

$(x + 3)(x + 1)$

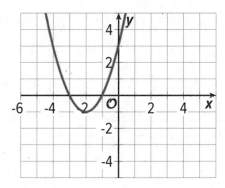

x-intercepts:

y-intercept:

Function g

$x^2 - 5x + 4$

$(x - 4)(x - 1)$

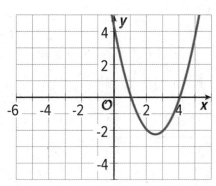

x-intercepts:

y-intercept:

Function h

$x^2 - 9$

$(x - 3)(x + 3)$

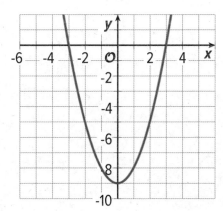

x-intercepts:

y-intercept:

Function i

$x^2 - 5x$

$x(x - 5)$

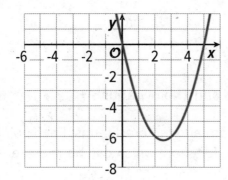

x-intercepts:

y-intercept:

Function j

$5x - x^2$

$x(5 - x)$

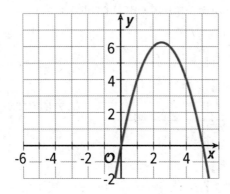

x-intercepts:

y-intercept:

Function k

$x^2 + 4x + 4$

$(x + 2)(x + 2)$

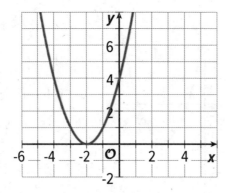

x-intercepts:

y-intercept:

NAME _____ DATE _____ PERIOD _____

2. What do you notice about the *x*-intercepts, the *y*-intercept, and the numbers in the expressions defining each function? Make a couple of observations.

3. Here is an expression that models function *p*, another quadratic function: $(x - 9)(x - 1)$. Predict the *x*-intercepts and the *y*-intercept of the graph that represent this function.

Are you ready for more?

Find the values of *a*, *p*, and *q* that will make $y = a(x - p)(x - q)$ be the equation represented by the graph.

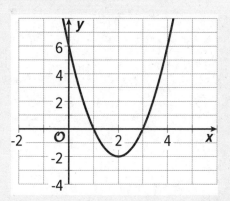

Different forms of quadratic functions can tell us interesting information about the function's graph. When a quadratic function is expressed in standard form, it can tell us the y-intercept of the graph representing the function. For example, the graph representing $y = x^2 - 5x + 7$ has its y-intercept (0, 7). This makes sense because the y-coordinate is the y-value when x is 0. Evaluating the expression at $x = 0$ gives $y = 0^2 - 5(0) + 7$, which equals 7.

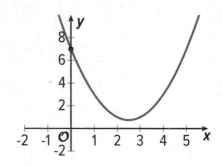

When a function is expressed in factored form, it can help us see the x-intercepts of its graph. Let's look at the functions f given by $f(x) = (x - 4)(x - 1)$ and g given by $g(x) = (x + 2)(x + 6)$.

If we graph $y = f(x)$, we see that the x-intercepts of the graph are (1, 0) and (4, 0). Notice that "1" and "4" also appear in $f(x) = (x - 4)(x - 1)$, and they are subtracted from x.

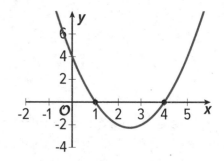

If we graph $y = g(x)$, we see that the x-intercepts are at (-2, 0) and (-6, 0). Notice that "2" and "6" are also in the equation $g(x) = (x + 2)(x + 6)$, but they are added to x.

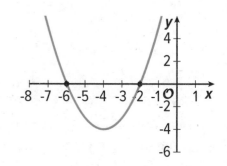

The connection between the factored form and the x-intercepts of the graph tells us about the zeros of the function (the input values that produce an output of 0). In the next lesson, we will further explore these connections between different forms of quadratic expressions and the graphs representing them.

NAME _____ DATE _____ PERIOD _____

Practice

Graphs of Functions in Standard and Factored Forms

1. A quadratic function f is defined by $f(x) = (x - 7)(x + 3)$.

   a. Without graphing, identify the x-intercepts of the graph of f. Explain how you know.

   b. Expand $(x - 7)(x + 3)$ and use the expanded form to identify the y-intercept of the graph of f.

2. What are the x-intercepts of the graph of the function defined by $(x - 2)(2x + 1)$?

   (A.) $(2, 0)$ and $(-1, 0)$

   (B.) $(2, 0)$ and $\left(-\frac{1}{2}, 0\right)$

   (C.) $(-2, 0)$ and $(1, 0)$

   (D.) $(-2, 0)$ and $\left(\frac{1}{2}, 0\right)$

3. Here is a graph that represents a quadratic function.

   Which expression could define this function?

   (A.) $(x + 3)(x + 1)$

   (B.) $(x + 3)(x - 1)$

   (C.) $(x - 3)(x + 1)$

   (D.) $(x - 3)(x - 1)$

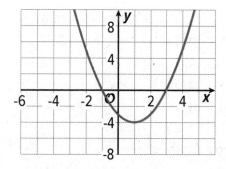

4. Respond to each question.

    a. What is the *y*-intercept of the graph of the equation

      $y = x^2 - 5x + 4$?

    b. An equivalent way to write this equation is $y = (x - 4)(x - 1)$.
       What are the *x*-intercepts of this equation's graph?

5. Noah said that if we graph $y = (x - 1)(x + 6)$, the *x*-intercepts will be at (1, 0) and (-6, 0). Explain how you can determine, without graphing, whether Noah is correct.

6. A company sells a video game. If the price of the game in dollars is *p* the company estimates that it will sell $20{,}000 - 500p$ games.

Which expression represents the revenue in dollars from selling games if the game is priced at *p* dollars? **(Lesson 6-7)**

   (**A.**) $(20{,}000 - 500p) + p$

   (**B.**) $(20{,}000 - 500p) - p$

   (**C.**) $\dfrac{20{,}000 - 500p}{p}$

   (**D.**) $(20{,}000 - 500p) \cdot p$

7. Write each quadratic expression in standard form. Draw a diagram if needed. **(Lesson 6-9)**

    a. $(x - 3)(x - 6)$

    b. $(x - 4)^2$

    c. $(2x + 3)(x - 4)$

    d. $(4x - 1)(3x - 7)$

NAME _____ DATE _____ PERIOD _____

8. Consider the expression $(5 + x)(6 - x)$. (Lesson 6-9)

   a. Is the expression equivalent to $x^2 + x + 30$? Explain how you know.

   b. Is the expression $30 + x - x^2$ in standard form? Explain how you know.

9. Here are graphs of the functions f and g given by
$f(x) = 100 \cdot \left(\frac{3}{5}\right)^x$ and $g(x) = 100 \cdot \left(\frac{2}{5}\right)^x$. (Lesson 5-12)

Which graph corresponds to f and which graph corresponds to g? Explain how you know.

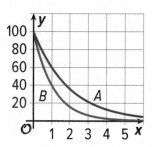

10. Here are graphs of two functions f and g.

An equation defining f is $f(x) = 100 \cdot 2^x$. (Lesson 5-13)

Which of these could be an equation defining the function g?

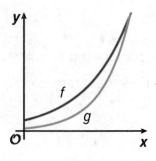

(A) $g(x) = 25 \cdot 3^x$

(B) $g(x) = 50 \cdot (1.5)^x$

(C) $g(x) = 100 \cdot 3^x$

(D) $g(x) = 200 \cdot (1.5)^x$

Lesson 6-11

Graphing from the Factored Form

NAME _____ DATE _____ PERIOD _____

Learning Goal Let's graph some quadratic functions in factored form.

Warm Up
11.1 Finding Coordinates

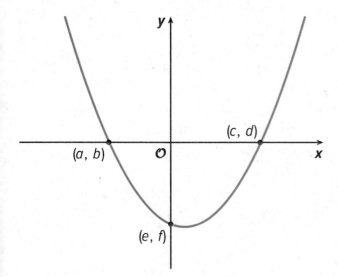

Here is a graph of a function w defined by $w(x) = (x + 1.6)(x - 2)$. Three points on the graph are labeled.

Find the values of $a, b, c, d, e,$ and f. Be prepared to explain your reasoning.

Consider two functions defined by $f(x) = x(x + 4)$ and $g(x) = x(x - 4)$.

1. Complete the table of values for each function. Then, determine the *x*-intercepts and vertex of each graph. Be prepared to explain how you know.

x	f(x)
-5	5
-4	
-3	
-2	-4
-1	-3
0	
1	
2	
3	
4	32
5	

x-intercepts:

Vertex:

x	g(x)
-5	45
-4	
-3	
-2	12
-1	5
0	
1	
2	
3	-3
4	
5	

x-intercepts:

Vertex:

NAME _____ DATE _____ PERIOD _____

2. Plot the points from the tables on the same coordinate plane. (Consider using different colors or markings for each set of points so you can tell them apart.) Then, make a couple of observations about how the two graphs compare.

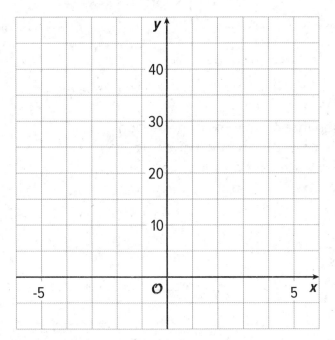

 Activity

11.3 What Do We Need to Sketch a Graph?

1. The functions f, g, and h are given. Predict the x-intercepts and the x-coordinate of the vertex of each function.

Equation	x-Intercepts	x-Coordinate of the Vertex
$f(x) = (x + 3)(x - 5)$		
$g(x) = 2x(x - 3)$		
$h(x) = (x + 4)(4 - x)$		

2. Use graphing technology to graph the functions f, g, and h. Use the graphs to check your predictions.

3. Without using technology, sketch a graph that represents the equation $y = (x - 7)(x + 11)$ and that shows the x-intercepts and the vertex. Think about how to find the y-coordinate of the vertex. Be prepared to explain your reasoning.

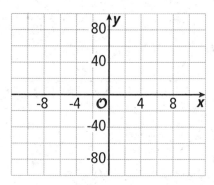

The quadratic function f is given by $f(x) = x^2 + 2x + 6$.

1. Find $f(-2)$ and $f(0)$.

2. What is the x-coordinate of the vertex of the graph of this quadratic function?

3. Does the graph have any x-intercepts? Explain or show how you know.

NAME _____ DATE _____ PERIOD _____

Summary
Graphing from the Factored Form

The function f given by $f(x) = (x + 1)(x - 3)$ is written in factored form. Recall that this form is helpful for finding the zeros of the function (where the function has the value 0) and telling us the x-intercepts on the graph representing the function.

Here is a graph representing f. It shows 2 x-intercepts at $x = -1$ and $x = 3$.

If we use -1 and 3 as inputs to f, what are the outputs?

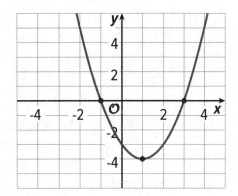

- $f(-1) = (-1 + 1)(-1 - 3) = (0)(-4) = 0$
- $f(3) = (3 + 1)(3 - 3) = (4)(0) = 0$

Because the inputs -1 and 3 produce an output of 0, they are the zeros of the function f. And because both x values have 0 for their y value, they also give us the x-intercepts of the graph (the points where the graph crosses the x-axis, which always have a y-coordinate of 0). So, the zeros of a function have the same values as the x-coordinates of the x-intercepts of the graph of the function.

The factored form can also help us identify the vertex of the graph, which is the point where the function reaches its minimum value. Notice that the x-coordinate of the vertex is 1, and that 1 is halfway between -1 and 3. Once we know the x-coordinate of the vertex, we can find the y-coordinate by evaluating the function: $f(1) = (1 + 1)(1 - 3) = 2(-2) = -4$. So the vertex is at $(1, -4)$.

When a quadratic function is in standard form, the y-intercept is clear: its y-coordinate is the constant term c in $ax^2 + bx + c$. To find the y-intercept from factored form, we can evaluate the function at $x = 0$, because the y-intercept is the point where the graph has an input value of 0.
$f(0) = (0 + 1)(0 - 3) = (1)(-3) = -3$.

Practice

Graphing from the Factored Form

1. Select **all** true statements about the graph that represents $y = 2x(x - 11)$.

   (A.) Its x-intercepts are at $(-2, 0)$ and $(11, 0)$.

   (B.) Its x-intercepts are at $(0, 0)$ and $(11, 0)$.

   (C.) Its x-intercepts are at $(2, 0)$ and $(-11, 0)$.

   (D.) It has only one x-intercept.

   (E.) The x-coordinate of its vertex is -4.5.

   (F.) The x-coordinate of its vertex is 11.

   (G.) The x-coordinate of its vertex is 4.5.

   (H.) The x-coordinate of its vertex is 5.5.

2. Select **all** equations whose graphs have a vertex with x-coordinate 2.

   (A.) $y = (x - 2)(x - 4)$

   (B.) $y = (x - 2)(x + 2)$

   (C.) $y = (x - 1)(x - 3)$

   (D.) $y = x(x + 4)$

   (E.) $y = x(x - 4)$

3. Determine the x-intercepts and the x-coordinate of the vertex of the graph that represents each equation.

Equation	x-Intercepts	x-Coordinate of the Vertex
$y = x(x - 2)$		
$y = (x - 4)(x + 5)$		
$y = -5x(3 - x)$		

NAME _____ DATE _____ PERIOD _____

4. Which one is the graph of the equation $y = (x - 3)(x + 5)$?

Graph A

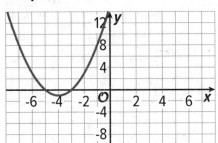

Graph B

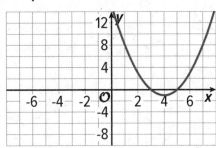

Graph C

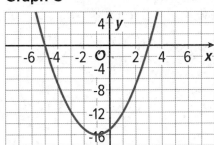

Graph D

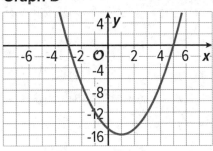

(A.) Graph A   (B.) Graph B   (C.) Graph C   (D.) Graph D

5. Respond to each question.

a. What are the x-intercepts of the graph of $y = (x - 2)(x - 4)$?

b. Find the coordinates of another point on the graph. Show your reasoning.

c. Sketch a graph of the equation $y = (x - 2)(x - 4)$.

6. A company sells calculators. If the price of the calculator in dollars is p, the company estimates that it will sell $10{,}000 - 120p$ calculators.

   Write an expression that represents the revenue in dollars from selling calculators if a calculator is priced at p dollars. (Lesson 6-7)

7. Is $(s + t)^2$ equivalent to $s^2 + 2st + t^2$? Explain or show your reasoning.
   (Lesson 6-8)

8. Tyler is shopping for a truck. He found two trucks that he likes. One truck sells for \$7,200. A slightly older truck sells for 15% less. How much does the older truck cost? (Lesson 5-14)

9. Here are graphs of two exponential functions, f and g.

   The function f is given by $f(x) = 100 \cdot 2^x$ while g is given by $g(x) = a \cdot b^x$.

   Based on the graphs of the functions, what can you conclude about a and b? (Lesson 5-13)

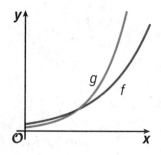

10. Suppose G takes a student's grade and gives a student's name as the output. Explain why G is not a function. (Lesson 4-2)

Lesson 6-12

Graphing the Standard Form (Part 1)

NAME _____ DATE _____ PERIOD _____

Learning Goal Let's see how the numbers in expressions like $-3x^2 + 4$ affect their graph.

Warm Up
12.1 Matching Graphs to Linear Equations

Graphs A, B, and C represent 3 linear equations: $y = 2x + 4$, $y = 3 - x$, and $y = 3x - 2$. Which graph corresponds to which equation? Explain your reasoning.

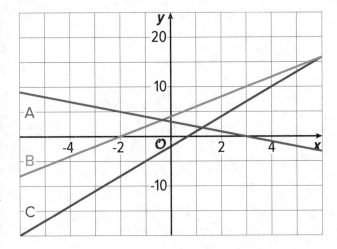

Using graphing technology, graph $y = x^2$, and then experiment with each of the following changes to the function. Record your observations (include sketches, if helpful).

1. Adding different constant terms to x^2 (for example: $x^2 + 5$, $x^2 + 10$, $x^2 - 3$, etc.)

2. Multiplying x^2 by different positive coefficients greater than 1 (for example: $3x^2$, $7.5x^2$, etc.)

3. Multiplying x^2 by different negative coefficients less than or equal to -1 (for example: $-x^2$, $-4x^2$, etc.)

4. Multiplying x^2 by different coefficients between -1 and 1 (for example: $\frac{1}{2}x^2$, $-0.25x^2$, etc.)

NAME _____ DATE _____ PERIOD _____

Are you ready for more?

Here are the graphs of three quadratic functions. What can you say about the coefficients of x^2 in the expressions that define f, g, and h? Can you identify them? How do they compare?

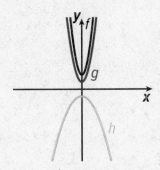

 ## Activity

12.3 What Do These Tables Reveal?

1. Respond to each question.

   a. Complete the table with values of $x^2 + 10$ and $x^2 - 3$ at different values of x. (You may also use a spreadsheet tool, if available.)

x	-3	-2	-1	0	1	2	3
x^2	9	4	1	0	1	4	9
$x^2 + 10$							
$x^2 - 3$							

b. Earlier, you observed the effects on the graph of adding or subtracting a constant term from x^2. Study the values in the table. Use them to explain why the graphs changed they way they did when a constant term is added or subtracted.

2. Respond to each question.

  a. Complete the table with values of $2x^2$, $\frac{1}{2}x^2$, and $-2x^2$ at different values of x. (You may also use a spreadsheet tool, if available.)

x	-3	-2	-1	0	1	2	3
x^2	9	4	1	0	1	4	9
$2x^2$							
$\frac{1}{2}x^2$							
$-2x^2$							

  b. You also observed the effects on the graph of multiplying x^2 by different coefficients. Study the values in the table. Use them to explain why the graphs changed they way they did when x^2 is multiplied by a number greater than 1, by a negative number less than or equal to -1, and by numbers between -1 and 1.

NAME _____ DATE _____ PERIOD _____

Activity

12.4 Card Sort: Representations of Quadratic Functions

Your teacher will give your group a set of cards. Each card contains a graph or an equation.

- Take turns with your partner to sort the cards into sets so that each set contains two equations and a graph that all represent the same quadratic function.

- For each set of cards that you put together, explain to your partner how you know they belong together.

- For each set that your partner puts together, listen carefully to their explanation. If you disagree, discuss your thinking and work to reach an agreement.

- Once all the cards are sorted and discussed, record the equivalent equations, sketch the corresponding graph, and write a brief note or explanation about why the representations were grouped together.

Standard form: Explanation:

Factored form:

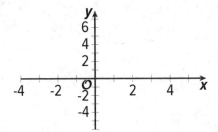

Standard form: Explanation:

Factored form:

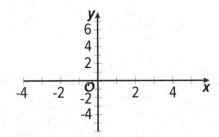

Standard form: Explanation:

Factored form:

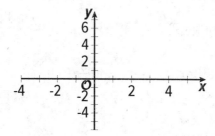

Standard form: Explanation:

Factored form:

NAME _____ DATE _____ PERIOD _____

Summary

Graphing the Standard Form (Part 1)

Remember that the graph representing any quadratic function is a shape called a *parabola*. People often say that a parabola "opens upward" when the lowest point on the graph is the vertex (where the graph changes direction), and "opens downward" when the highest point on the graph is the vertex. Each coefficient in a quadratic expression written in standard form $ax^2 + bx + c$ tells us something important about the graph that represents it.

The graph of $y = x^2$ is a parabola opening upward with vertex at $(0, 0)$. Adding a constant term 5 gives $y = x^2 + 5$ and raises the graph by 5 units. Subtracting 4 from x^2 gives $y = x^2 - 4$ and moves the graph 4 units down.

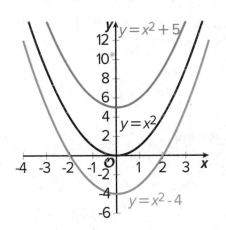

x	-3	-2	-1	0	1	2	3
x^2	9	4	1	0	1	4	9
$x^2 + 5$	14	9	6	5	6	9	14
$x^2 - 4$	5	0	-3	-4	-3	0	5

A table of values can help us see that adding 5 to x^2 increases all the output values of $y = x^2$ by 5, which explains why the graph moves up 5 units. Subtracting 4 from x^2 decreases all the output values of $y = x^2$ by 4, which explains why the graph shifts down by 4 units.

In general, the constant term of a quadratic expression in standard form influences the vertical position of the graph. An expression with no constant term (such as x^2 or $x^2 + 9x$) means that the constant term is 0, so the y-intercept of the graph is on the x-axis. It's not shifted up or down relative to the x-axis.

The coefficient of the squared term in a quadratic function also tells us something about its graph. The coefficient of the squared term in $y = x^2$ is 1. Its graph is a parabola that opens upward.

- Multiplying x^2 by a number greater than 1 makes the graph steeper, so the parabola is narrower than that representing x^2.

- Multiplying x^2 by a number less than 1 but greater than 0 makes the graph less steep, so the parabola is wider than that representing x^2.

- Multiplying x^2 by a number less than 0 makes the parabola open downward.

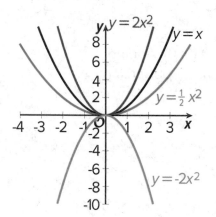

x	-3	-2	-1	0	1	2	3
x^2	9	4	1	0	1	4	9
$2x^2$	18	8	2	0	2	8	18
$-2x^2$	-18	-8	-2	0	-2	-8	-18

If we compare the output values of $2x^2$ and $-2x^2$, we see that they are opposites, which suggests that one graph would be a reflection of the other across the x-axis.

NAME _____ DATE _____ PERIOD _____

Practice
Graphing the Standard Form (Part 1)

1. Here are four graphs. Match each graph with a quadratic equation that it represents.

Graph A

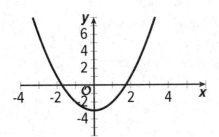

Graph B

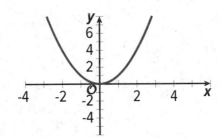

Graph C

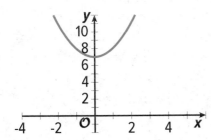

Graph D

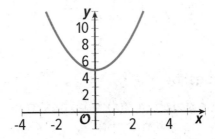

A. Graph A

B. Graph B

C. Graph C

D. Graph D

1. $y = x^2$

2. $y = x^2 + 5$

3. $y = x^2 + 7$

4. $y = x^2 - 3$

2. The two equations $y = (x + 2)(x + 3)$ and $y = x^2 + 5x + 6$ are equivalent.

   a. Which equation helps find the x-intercepts most efficiently?

   b. Which equation helps find the y-intercept most efficiently?

3. Here is a graph that represents $y = x^2$.

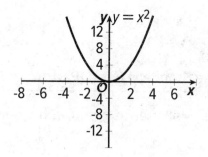

On the same coordinate plane, sketch and label the graph that represents each equation

a. $y = x^2 - 4$

b. $y = -x^2 + 5$

4. Select **all** equations whose graphs have a y-intercept with a positive y-coordinate.

(A.) $y = x^2 + 3x - 2$

(B.) $y = x^2 - 10x$

(C.) $y = (x - 1)^2$

(D.) $y = 5x^2 - 3x - 5$

(E.) $y = (x + 1)(x + 2)$

5. Respond to each question.

a. Describe how the graph of $A(x) = |x|$ has to be shifted to match the given graph. **(Lesson 4-14)**

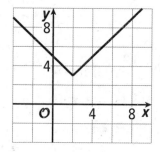

b. Write an equation for the function represented by the graph.

6. Here is a graph of the function g given by $g(x) = a \cdot b^x$. **(Lesson 5-13)**

What can you say about the value of b? Explain how you know.

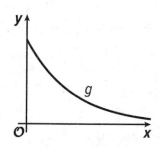

NAME _____ DATE _____ PERIOD _____

7. Respond to each question. **(Lesson 6-11)**

    a. What are the x-intercepts of the graph that represents $y = (x + 1)(x + 5)$? Explain how you know.

    b. What is the x-coordinate of the vertex of the graph that represents $y = (x + 1)(x + 5)$? Explain how you know.

    c. Find the y-coordinate of the vertex. Show your reasoning.

    d. Sketch a graph of $y = (x + 1)(x + 5)$.

8. Determine the x-intercepts, the vertex, and the y-intercept of the graph of each equation. **(Lesson 6-11)**

Equation	x-intercepts	vertex	y-intercept
$y = (x - 5)(x - 3)$			
$y = 2x(8 - x)$			

9. Equal amounts of money were invested in stock A and stock B. In the first year, stock A increased in value by 20%, and stock B decreased by 20%. In the second year, stock A decreased in value by 20%, and stock B increased by 20%. (Lesson 5-15)

Was one stock a better investment than the other? Explain your reasoning.

Lesson 6-13

Graphing the Standard Form (Part 2)

NAME _____ DATE _____ PERIOD _____

Learning Goal Let's change some other parts of a quadratic expression and see how they affect the graph.

Warm Up
13.1 Equivalent Expressions

1. Complete each row with an equivalent expression in standard form or factored form.

Standard Form	Factored Form
x^2	
	$x(x + 9)$
$x^2 - 18x$	
	$x(6 - x)$
$-x^2 + 10x$	
	$-x(x + 2.75)$

2. What do the quadratic expressions in each column have in common (besides the fact that everything in the left column is in standard form and everything in the other column is in factored form)? Be prepared to share your observations.

1. Using graphing technology:

   a. Graph $y = x^2$, and then experiment with adding different linear terms (for example, $x^2 + 4x$, $x^2 + 20x$, $x^2 - 50x$). Record your observations.

   b. Graph $y = -x^2$, and then experiment with adding different linear terms. Record your observations.

2. Use your observations to help you complete the table without graphing the equations.

Equation	x-Intercepts	x-Coordinate of Vertex
$y = x^2 + 6x$		
$y = x^2 - 10x$		
$y = -x^2 + 50x$		
$y = -x^2 - 36x$		

3. Some quadratic expressions have no linear terms. Find the x-intercepts and the x-coordinate of the vertex of the graph representing each equation. (Note it is possible for the graph to not intersect the x-axis.) If you get stuck, try graphing the equations.

   a. $y = x^2 - 25$

   b. $y = x^2 + 16$

NAME _____ DATE _____ PERIOD _____

Activity

13.3 Writing Equations to Match Graphs

Use graphing technology to graph a function that matches each given graph.
Make sure your graph goes through all 3 points shown!

A

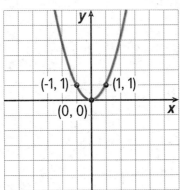

Equation:

B

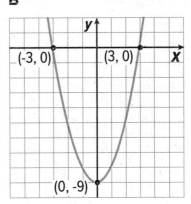

Equation:

C

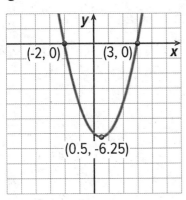

Equation:

D

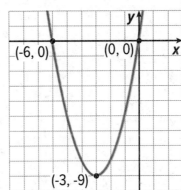

Equation:

E

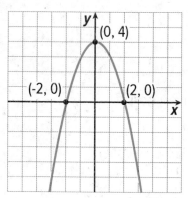

(0, 4)

(-2, 0)   (2, 0)

Equation: _____

F

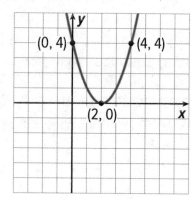

(0, 4)   (4, 4)

(2, 0)

Equation: _____

G

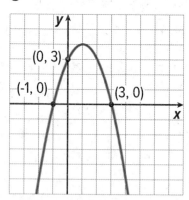

(0, 3)

(-1, 0)   (3, 0)

Equation: _____

H

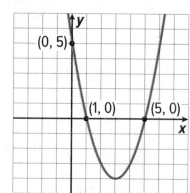

(0, 5)

(1, 0)   (5, 0)

Equation: _____

I

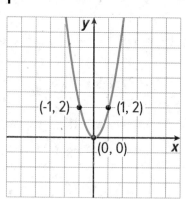

(-1, 2)   (1, 2)

(0, 0)

Equation: _____

J

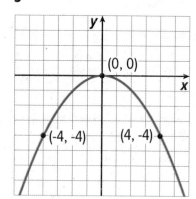

(0, 0)

(-4, -4)   (4, -4)

Equation: _____

NAME _____ DATE _____ PERIOD _____

Summary
Graphing the Standard Form (Part 2)

In an earlier lesson, we saw that a quadratic function written in standard form $ax^2 + bx + c$ can tell us some things about the graph that represents it. The coefficient a can tell us whether the graph of the function opens upward or downward, and also gives us information about whether it is narrow or wide. The constant term c can tell us about its vertical position.

Recall that the graph representing $y = x^2$ is an upward-opening parabola with the vertex at (0, 0). The vertex is also the x-intercept and the y-intercept. Suppose we add 6 to the squared term: $y = x^2 + 6$. Adding a 6 shifts the graph upwards, so the vertex is at (0, 6). The vertex is the y-intercept and the graph is centered on the y-axis.

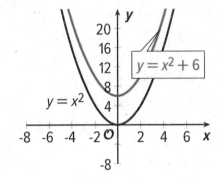

What can the linear term bx tell us about the graph representing a quadratic function? The linear term has a somewhat mysterious effect on the graph of a quadratic function. The graph seems to shift both horizontally and vertically. When we add bx (where b is not 0) to x^2, the graph of $y = x^2 + bx$ is no longer centered on the y-axis.

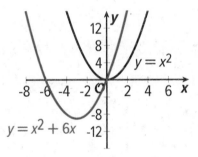

Suppose we add $6x$ to the squared term: $y = x^2 + 6x$. Writing the $x^2 + 6x$ in factored form as $x(x + 6)$ gives us the zeros of the function, 0 and -6. Adding the term $6x$ seems to shift the graph to the left and down and the x-intercepts are now (-6, 0) and (0, 0). The vertex is no longer the y-intercept and the graph is no longer centered on the y-axis.

What if we add -6x to x^2? $x^2 - 6x$ can be rewritten as $x(x - 6)$, which tells us the zeros: 0 and 6. Adding a negative linear term to a squared term seems to shift the graph to the right and down. The x-intercepts are now (0, 0) and (6, 0). The vertex is no longer the y-intercept and the graph is not centered on the y-axis.

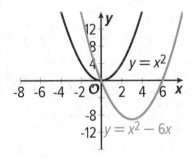

Practice

Graphing the Standard Form (Part 2)

1. Here are four graphs. Match each graph with the quadratic equation that it represents.

Graph A

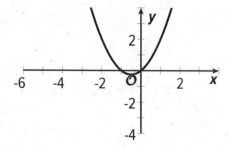

Graph B

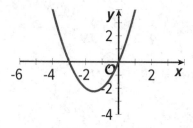

Graph C

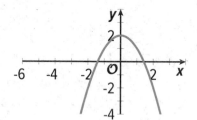

Graph D

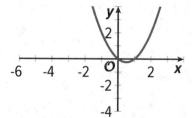

A. Graph A

B. Graph B

C. Graph C

D. Graph D

1. $y = x^2 + x$

2. $y = -x^2 + 2$

3. $y = x^2 - x$

4. $y = x^2 + 3x$

2. Complete the table without graphing the equations.

Equation	x-Intercepts	x-Coordinate of Vertex
$y = x^2 + 12x$		
$y = x^2 - 3x$		
$y = -x^2 + 16x$		
$y = -x^2 - 24x$		

NAME _____ DATE _____ PERIOD _____

3. Here is a graph that represents $y = x^2$.

    a. Describe what would happen to the graph if the original equation were changed to $y = x^2 - 6x$. Predict the x- and y-intercepts of the graph and the quadrant where the vertex is located.

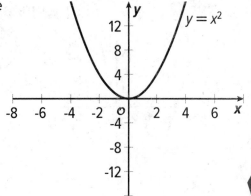

    b. Sketch the graph of the equation $y = x^2 - 6x$ on the same coordinate plane as $y = x^2$.

4. Select **all** equations whose graph opens upward.

    Ⓐ $y = -x^2 + 9x$

    Ⓑ $y = 10x - 5x^2$

    Ⓒ $y = (2x - 1)^2$

    Ⓓ $y = (1 - x)(2 + x)$

    Ⓔ $y = x^2 - 8x - 7$

5. *Technology required.* Write an equation for a function that can be represented by each given graph. Then, use graphing technology to check each equation you wrote.

Graph 1

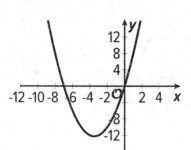

Graph 2

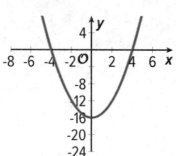

Graph 3

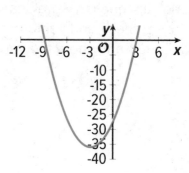

6. Match each quadratic expression that is written as a product with an equivalent expression that is expanded. (Lesson 6-8)

   A. $(x + 3)(x + 4)$

   B. $(x + 3)(x + 7)$

   C. $(3x + 4)(x + 3)$

   D. $(x + 7)(3x + 1)$

   1. $x^2 + 10x + 21$

   2. $3x^2 + 13x + 12$

   3. $3x^2 + 22x + 7$

   4. $x^2 + 7x + 12$

7. When buying a home, many mortgage companies require a down payment of 20% of the price of the house. What is the down payment on a $125,000 home? (Lesson 5-14)

8. A bank loans $4,000 to a customer at a $9\frac{1}{2}$% annual interest rate. (Lesson 5-15)

   Write an expression to represent how much the customer will owe, in dollars, after 5 years without payment.

Lesson 6-14

Graphs That Represent Situations

NAME _____ DATE _____ PERIOD _____

Learning Goal Let's examine graphs that represent the paths of objects being launched in the air.

Warm Up
14.1 A Jumping Frog

The height in inches of a frog's jump is modeled by the equation $h(t) = 60t - 75t^2$ where the time, t, after it jumped is measured in seconds.

1. Find $h(0)$ and $h(0.8)$. What do these values mean in terms of the frog's jump?

2. How much time after it jumped did the frog reach the maximum height? Explain how you know.

Activity

14.2 A Catapulted Pumpkin

The equation $h = 2 + 23.7t - 4.9t^2$ represents the height of a pumpkin that is catapulted up in the air as a function of time, t, in seconds. The height is measured in meters above ground. The pumpkin is shot up at a vertical velocity of 23.7 meters per second.

1. Without writing anything down, consider these questions:

   - What do you think the 2 in the equation tells us in this situation? What about the $-4.9t^2$?

   - If we graph the equation, will the graph open upward or downward? Why?

   - Where do you think the vertical intercept would be?

   - What about the horizontal intercepts?

2. Graph the equation using graphing technology.

NAME _____ DATE _____ PERIOD _____

3. Identify the vertical and horizontal intercepts, and the vertex of the graph. Explain what each point means in this situation.

Are you ready for more?

What approximate vertical velocity would this pumpkin need for it stay in the air for about 10 seconds? (Assume that it is still shot from 2 meters in the air and that the effect of gravity pulling it down is the same.)

Here is a graph that represents the height of a baseball, *h*, in feet as a function of time, *t*, in seconds after it was hit by Player A.

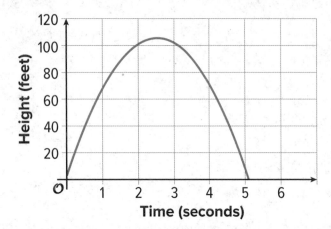

The function *g* defined by $g(t) = (-16t - 1)(t - 4)$ also represents the height in feet of a baseball *t* seconds after it was hit by Player B. Without graphing function *g*, answer the following questions and explain or show how you know.

1. Which player's baseball stayed in flight longer?

2. Which player's baseball reached a greater maximum height?

3. How can you find the height at which each baseball was hit?

NAME _____ DATE _____ PERIOD _____

Activity
14.4 Info Gap: Rocket Math

Your teacher will give you either a problem card or a data card. Do not show or read your card to your partner.

If your teacher gives you the data card:

1. Silently read the information on your card.

2. Ask your partner "What specific information do you need?" and wait for your partner to ask for information. Only give information that is on your card. (Do not figure out anything for your partner!)

3. Before telling your partner the information, ask "Why do you need to know (that piece of information)?"

4. Read the problem card, and solve the problem independently.

5. Share the data card, and discuss your reasoning.

If your teacher gives you the problem card:

1. Silently read your card and think about what information you need to answer the question.

2. Ask your partner for the specific information that you need.

3. Explain to your partner how you are using the information to solve the problem.

4. When you have enough information, share the problem card with your partner, and solve the problem independently.

5. Read the data card, and discuss your reasoning.

Pause here so your teacher can review your work. Ask your teacher for a new set of cards and repeat the activity, trading roles with your partner.

Let's say a tennis ball is hit straight up in the air, and its height in feet above the ground is modeled by the equation $f(t) = 4 + 12t - 16t^2$, where t represents the time in seconds after the ball is hit. Here is a graph that represents the function, from the time the tennis ball was hit until the time it reached the ground.

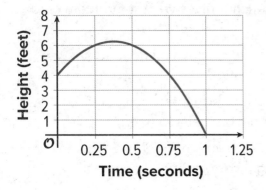

In the graph, we can see some information we already know, and some new information:

- The 4 in the equation means the graph of the function intersects the vertical axis at 4. It shows that the tennis ball was 4 feet off the ground at $t = 0$, when it was hit.

- The horizontal intercept is (1, 0). It tells us that the tennis ball hits the ground 1 second after it was hit.

- The vertex of the graph is at approximately (0.4, 6.3). This means that about 0.4 second after the ball was hit, it reached the maximum height of about 6.3 feet.

The equation can be written in factored form as $f(t) = (-16t - 4)(t - 1)$. From this form, we can see that the zeros of the function are $t = 1$ and $t = -\frac{1}{4}$. The negative zero, $-\frac{1}{4}$, is not meaningful in this situation, because the time before the ball was hit is irrelevant.

NAME _____ DATE _____ PERIOD _____

Practice
Graphs That Represent Situations

1. Here are graphs of functions f and g.

   Each represents the height of an object being launched into the air as a function of time.

   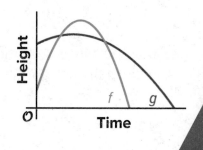

   a. Which object was launched from a higher point?

   b. Which object reached a higher point?

   c. Which object was launched with the higher upward velocity?

   d. Which object landed last?

2. *Technology required.* The function h given by $h(t) = (1 - t)(8 + 16t)$ models the height of a ball in feet, t seconds after it was thrown.

   a. Find the zeros of the function. Show or explain your reasoning.

   b. What do the zeros tell us in this situation? Are both zeros meaningful?

c. From what height is the ball thrown? Explain your reasoning.

d. About when does the ball reach its highest point, and about how high does the ball go? Show or explain your reasoning.

3. The height in feet of a thrown football is modeled by the equation $f(t) = 6 + 30t - 16t^2$, where time t is measured in seconds.

   a. What does the constant 6 mean in this situation?

   b. What does the $30t$ mean in this situation?

   c. How do you think the squared term $-16t^2$ affects the value of the function f? What does this term reveal about the situation?

4. The height in feet of an arrow is modeled by the equation $h(t) = (1 + 2t)(18 - 8t)$, where t is seconds after the arrow is shot.

   a. When does the arrow hit the ground? Explain or show your reasoning.

   b. From what height is the arrow shot? Explain or show your reasoning.

NAME _____ DATE _____ PERIOD _____

5. Two objects are launched into the air.

   • The height, in feet, of Object A is given by the equation $f(t) = 4 + 32t - 16t^2$.

   • The height, in feet, of the Object B is given by the equation $g(t) = 2.5 + 40t - 16t^2$. In both functions, t is seconds after launch.

   a. Which object was launched from a greater height? Explain how you know.

   b. Which object was launched with a greater upward velocity? Explain how you know.

6. Respond to each question. **(Lesson 6-10)**

   a. Predict the x- and y-intercepts of the graph of the quadratic function defined by the expression $(x + 6)(x - 6)$. Explain how you made your predictions.

   b. *Technology required.* Check your predictions by graphing $y = (x + 6)(x - 6)$.

7. *Technology required.* A student needs to get a loan of $12,000 for the first year of college. Bank A has an annual interest rate of 5.75%, Bank B has an annual interest rate of 7.81%, and Bank C has an annual rate of 4.45%. (Lesson 5-15)

a. If we graph the amount owed for each loan as a function of years without payment, predict what the three graphs would look like. Describe or sketch your prediction.

b. Use graphing technology to plot the graph of each loan balance.

c. Based on your graph, how much would the student owe for each loan when they graduate from college in four years?

d. Based on your graph, if no payments are made, how much would the student owe for each loan after 10 years?

8. *Technology required.* The functions f and g are given by $f(x) = 13x + 6$ and $g(x) = 0.1 \cdot (1.4)^x$. (Lesson 5-19)

a. Which function eventually grows faster, f or g? Explain how you know.

b. Use graphing technology to decide when the graphs of f and g meet.

Lesson 6-15

Vertex Form

NAME _____ DATE _____ PERIOD _____

Learning Goal Let's find out about the vertex form.

 Warm Up

15.1 Notice and Wonder: Two Sets of Equations

What do you notice? What do you wonder?

Set 1:

$f(x) = x^2 + 4x$

$g(x) = x(x + 4)$

$h(x) = (x + 2)^2 - 4$

Set 2:

$p(x) = -x^2 + 6x - 5$

$q(x) = (5 - x)(x - 1)$

$r(x) = -1(x - 3)^2 + 4$

Activity

15.2 A Whole New Form

Here are two sets of equations for quadratic functions you saw earlier. In each set, the expressions that define the output are equivalent.

Set 1:

$f(x) = x^2 + 4x$

$g(x) = x(x + 4)$

$h(x) = (x + 2)^2 - 4$

Set 2:

$p(x) = -x^2 + 6x - 5$

$q(x) = (5 - x)(x - 1)$

$r(x) = -1(x - 3)^2 + 4$

The expression that defines h is written in **vertex form**. We can show that it is equivalent to the expression defining f by expanding the expression:

$$(x + 2)^2 - 4 = (x + 2)(x + 2) - 4$$
$$= x^2 + 2x + 2x + 4 - 4$$
$$= x^2 + 4x$$

1. Show that the expressions defining r and p are equivalent.

2. Here are graphs representing the quadratic functions. Why do you think expressions such as those defining h and r are said to be written in vertex form?

Graph of h

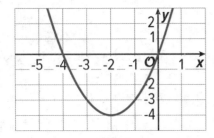

Graph of r

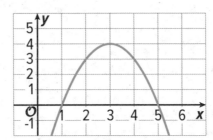

NAME _____ DATE _____ PERIOD _____

 ## Activity

15.3 Playing with Parameters

1. Using graphing technology, graph $y = x^2$. Then, add different numbers to x before it is squared (for example, $y = (x + 4)^2$, $y = (x - 3)^2$) and observe how the graph changes. Record your observations.

2. Graph $y = (x - 1)^2$. Then, experiment with each of the following changes to the function and see how they affect the graph and the vertex:

   a. Adding different constant terms to $(x - 1)^2$ (for example: $(x - 1)^2 + 5$, $(x - 1)^2 - 9$).

   b. Multiplying $(x - 1)^2$ by different coefficients (for example: $y = 3(x - 1)^2$, $y = -2(x - 1)^2$).

3. Without graphing, predict the coordinates of the vertex of the graphs of these quadratic functions, and predict whether the graph opens up or opens down. Ignore the last row until the next question.

Equations	Coordinates of Vertex	Graph Opens Up or Down?
$y = (x + 10)^2$		
$y = (x - 4)^2 + 8$		
$y = -(x - 4)^2 + 8$		
$y = x^2 - 7$		
$y = \frac{1}{2}(x + 3)^2 - 5$		
$y = -(x + 100)^2 + 50$		
$y = a(x + m)^2 + n$		

4. Use graphing technology to check your predictions. If they are incorrect, revise them. Then, complete the last row of the table.

Are you ready for more?

1. What is the vertex of this graph?

2. Find a quadratic equation whose graph has the same vertex and adjust it, if needed, so that it has the graph provided.

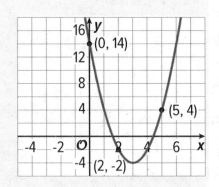

NAME _____ DATE _____ PERIOD _____

Summary
Vertex Form

Sometimes the expressions that define quadratic functions are written in **vertex form**. For example, if the function *f* is defined by $(x - 3)^2 + 4$, which is in vertex form, we can write $f(x) = (x - 3)^2 + 4$ and draw this graph to represent *f*.

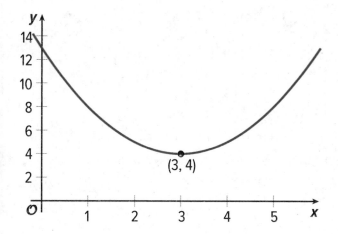

The vertex form can tell us about the coordinates of the vertex of the graph of a quadratic function. The expression $(x - 3)^2$ reveals that the vertex has *x*-coordinate 3, and the constant term of 4 reveals its *y*-coordinate. Here the vertex represents the minimum value of the function *f*, and its graph opens upward.

In general, a quadratic function expressed in vertex form is written as:

$$y = a(x - h)^2 + k$$

The vertex of its graph is at (h, k). The graph of the quadratic function opens upward when the coefficient *a* is positive and opens downward when *a* is negative.

In future lessons, we will explore further how *a*, *h*, and *k* affect the graph of a quadratic function.

Glossary

vertex form (of a quadratic expression)

Practice
Vertex Form

1. Select **all** of the quadratic expressions in vertex form.

   (A.) $(x - 2)^2 + 1$ (D.) $(x + 3)^2$

   (B.) $x^2 - 4$ (E.) $(x - 4)^2 + 6$

   (C.) $x(x + 1)$

2. Here are two equations. One defines function m and the other defines function p.

$$m(x) = x(x + 6) \qquad\qquad p(x) = (x + 3)^2 - 9$$

   a. Show that the expressions defining m and p are equivalent.

   b. What is the vertex of the graph of m? Explain how you know.

   c. What are the x-intercepts of the graph of p? Explain how you know.

3. Which equation is represented by the graph?

   (A.) $y = (x - 1)^2 + 3$ (C.) $y = -(x + 3)^2 - 1$

   (B.) $y = (x - 3)^2 + 1$ (D.) $y = -(x - 3)^2 + 1$

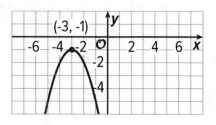

4. For each equation, write the coordinates of the vertex of the graph that represents the equation.

   a. $y = (x - 3)^2 + 5$ d. $y = x^2 - 1$

   b. $y = (x + 7)^2 + 3$ e. $y = 2(x + 1)^2 - 5$

   c. $y = (x - 4)^2$ f. $y = -2(x + 1)^2 - 5$

NAME _____ DATE _____ PERIOD _____

5. For each function, write the coordinates of the vertex of its graph and tell whether the graph opens up or down.

Function	Coordinates of Vertex	Graph Opens Up or Down?
$f(x) = (x - 4)^2 - 5$		
$g(x) = -x^2 + 5$		
$h(x) = 2(x + 1)^2 - 4$		

6. Here is a graph that represents $y = x^2$. (Lesson 6-12)

a. Describe what would happen to the graph if the original equation were modified as follows:

i. $y = -x^2$

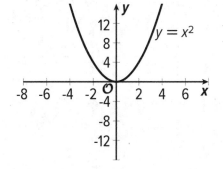

ii. $y = 3x^2$

iii. $y = x^2 + 6$

b. Sketch the graph of the equation $y = -3x^2 + 6$ on the same coordinate plane as $y = x^2$.

7. Noah is going to put \$2,000 in a savings account. He plans on putting the money in an account and leaving it there for 5 years. He can put the money in an account that pays 1% interest monthly, an account that pays 6% interest every six months, or an account that pays 12% interest annually. Which account will give him the most money in his account at the end of the 5 years? (Lesson 5-16)

8. Here are four graphs. Match each graph with a quadratic equation that it represents. **(Lesson 6-12)**

Graph A

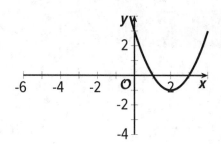

Graph B

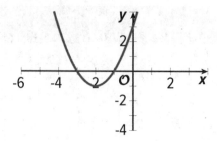

Graph C

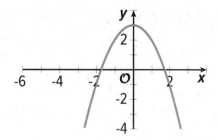

Graph D

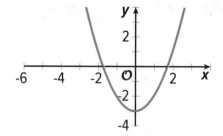

A. Graph A

B. Graph B

C. Graph C

D. Graph D

1. $y = -x^2 + 3$

2. $y = (x + 1)(x + 3)$

3. $y = x^2 - 3$

4. $y = (x - 1)(x - 3)$

9. The table shows some input and output values of function f. Describe a possible rule for the function by using words or by writing an equation. **(Lesson 4-5)**

x	$f(x)$
-3	-8
0	-2
4	6
10	18

Lesson 6-16

Graphing from the Vertex Form

NAME _____ DATE _____ PERIOD _____

Learning Goal Let's graph equations in vertex form.

 ## Warm Up

16.1 Which Form to Use?

Expressions in different forms can be used to define the same function. Here are three ways to define a function f.

$f(x) = x^2 - 4x + 3$ (standard form)

$f(x) = (x - 3)(x - 1)$ (factored form)

$f(x) = (x - 2)^2 - 1$ (vertex form)

Which form would you use if you want to find the following features of the graph of f? Be prepared to explain your reasoning.

1. the x-intercepts

2. the vertex

3. the y-intercept

Here are two equations that define quadratic functions.

$$p(x) = -(x - 4)^2 + 10$$

$$q(x) = \frac{1}{2}(x - 4)^2 + 10$$

1. The graph of p passes through (0, -6) and (4, 10), as shown on the coordinate plane.

    Find the coordinates of another point on the graph of p. Explain or show your reasoning. Then, use the points to sketch and label the graph.

2. On the same coordinate plane, identify the vertex and two other points that are on the graph of q. Explain or show your reasoning. Sketch and label the graph of q.

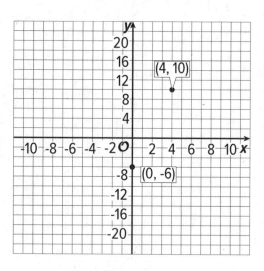

NAME _____ DATE _____ PERIOD _____

3. Priya says, "Once I know the vertex is (4, 10), I can find out, without graphing, whether the vertex is the maximum or the minimum of function p. I would just compare the coordinates of the vertex with the coordinates of a point on either side of it."

Complete the table and then explain how Priya might have reasoned about whether the vertex is the minimum or maximum.

x	3	4	5
$p(x)$		10	

Are you ready for more?

1. Write an equation for a quadratic function whose graph has the vertex at (2, 3) and contains the point (0, -5).

2. Sketch a graph of your function.

Your teacher will give you a set of cards. Each card contains an equation or a graph that represents a quadratic function. Take turns matching each equation to a graph that represents the same function.

- For each pair of cards that you match, explain to your partner how you know they belong together.

- For each pair of cards that your partner matches, listen carefully to their explanation. If you disagree, discuss your thinking and work to reach an agreement.

- Once all the cards are matched, record the equation, the label and a sketch of the corresponding graph, and write a brief note or explanation about how you knew they were a match.

Equation:

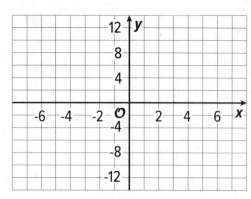

Explanation:

Equation:

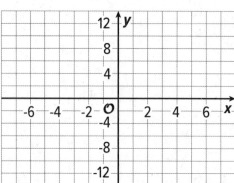

Explanation:

Equation:

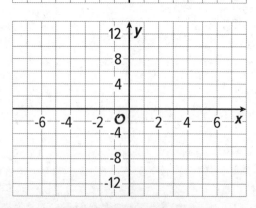

Explanation:

NAME _____ DATE _____ PERIOD _____

Equation:

Explanation:

Equation:

Explanation:

Equation:

Explanation:

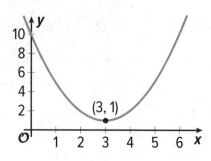

We saw that vertex form is especially helpful for finding the vertex of a graph of a quadratic function. For example, we can tell that the function p given by $p(x) = (x - 3)^2 + 1$ has a vertex at (3, 1).

We also noticed that, when the squared expression $(x - 3)^2$ has a positive coefficient, the graph opens upward. This means that the vertex (3, 1) represents the minimum function value.

But why does the function p take on its minimum value when x is 3?

Here is one way to explain it: When $x = 3$, the squared term $(x - 3)^2$ equals 0, as $(3 - 3)^2 = 0^2 = 0$. When x is any other value besides 3, the squared term $(x - 3)^2$ is a positive number greater than 0. (Squaring any number results in a positive number.) This means that the output when $x \neq 3$ will always be greater than the output when $x = 3$, so the function p has a minimum value at $x = 3$.

This table shows some values of the function for some values of x. Notice that the output is the least when $x = 3$ and it increases both as x increases and as it decreases.

x	0	1	2	3	4	5	6
$(x - 3)^2 + 1$	10	5	2	1	2	5	10

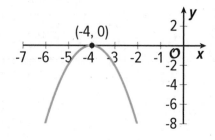

The squared term sometimes has a negative coefficient, for instance: $h(x) = -2(x + 4)^2$. The x value that makes $(x + 4)^2$ equal 0 is -4, because $(-4 + 4)^2 = 0^2 = 0$. Any other x value makes $(x + 4)^2$ greater than 0. But when $(x + 4)^2$ is multiplied by a negative number (-2), the resulting expression, $-2(x + 4)^2$, ends up being negative. This means that the output when $x \neq -4$ will always be less than the output when $x = -4$, so the function h has its maximum value when $x = -4$.

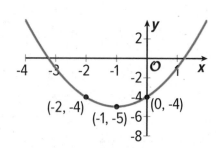

Remember that we can find the y-intercept of the graph representing any function we have seen. The y-coordinate of the y-intercept is the value of the function when $x = 0$. If g is defined by $g(x) = (x + 1)^2 - 5$, then the y-intercept is (0, -4) because $g(0) = (0 + 1)^2 - 5 = -4$. Its vertex is at (-1, -5).

Another point on the graph with the same y-coordinate is located the same horizontal distance from the vertex but on the other side.

NAME _____ DATE _____ PERIOD _____

Practice
Graphing from the Vertex Form

1. Which equation can be represented by a graph with a vertex at (1, 3)?

   (A.) $y = (x - 1)^2 + 3$

   (B.) $y = (x + 1)^2 + 3$

   (C.) $y = (x - 3)^2 + 1$

   (D.) $y = (x + 3)^2 + 1$

2. Respond to each question.

   a. Where is the vertex of the graph that represents $y = (x - 2)^2 - 8$?

   b. Where is the *y*-intercept? Explain how you know.

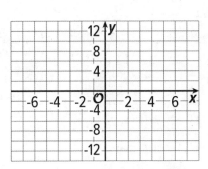

   c. Identify one other point on the graph of the equation. Explain or show how you know.

   d. Sketch a graph that represents the equation.

3. The function v is defined by $v(x) = \frac{1}{2}(x + 5)^2 - 7$.

   Without graphing, determine if the vertex of the graph representing v shows the minimum or maximum value of the function. Explain how you know.

4. Match each graph to an equation that represents it.

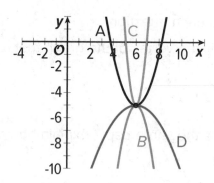

A. Graph A

B. Graph B

C. Graph C

D. Graph D

1. $y = -2(x - 6)^2 - 5$

2. $y = (x - 6)^2 - 5$

3. $y = 6(x - 6)^2 - 5$

4. $y = -\frac{1}{3}(x - 6)^2 - 5$

NAME _____ DATE _____ PERIOD _____

5. Here is a graph that represents $y = x^2$. (Lesson 6-12)

   a. Describe what would happen to the graph if the original equation was changed to:

   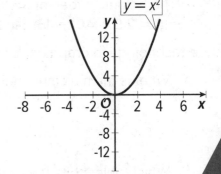

   I. $y = \frac{1}{2}x^2$

   II. $y = x^2 - 8$

   b. Graph the equation $y = \frac{1}{2}x^2 - 8$ on the same coordinate plane as $y = x^2$.

6. Clare throws a rock into the lake. The graph shows the rock's height above the water, in feet, as a function of time in seconds. (Lesson 6-14)

   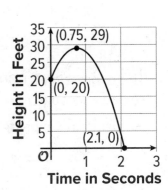

   Select **all** the statements that describe this situation.

   (A.) The vertex of the graph is (0.75, 29).

   (B.) The y-intercept of the graph is (2.1, 0).

   (C.) Clare just dropped the rock into the lake.

   (D.) The maximum height of the rock is about 20 feet.

   (E.) The rock hits the surface of the water after about 2.1 seconds.

   (F.) Clare tossed the rock up into the air from a point 20 feet above the water.

7. *Technology required.* Two objects are launched into the air. (Lesson 6-14)

- The height, in feet, of Object A is given by the equation
$f(t) = 4 + 32t - 16t^2$.

- The height, in feet, of the Object B is given by the equation
$g(t) = 2.5 + 40t - 16t^2$. In both functions, t is seconds after launch.

Use technology to graph each function in the same graphing window.

a. What is the maximum height of each object?

b. Which object hits the ground first? Explain how you know.

8. Andre thinks the vertex of the graph of the equation $y = (x + 2)^2 - 3$ is
(2, -3). Lin thinks the vertex is (-2, 3). Do you agree with either of them?
(Lesson 6-15)

9. The expression $2{,}000 \cdot (1.015^{12})^5$ represents the balance, in dollars, in a
savings account. (Lesson 5-17)

a. Using the expression, describe the interest rate paid on the account?

b. How many years has the account been accruing interest?

c. How much money was invested?

d. How much money is in the account now?

e. Write an equivalent expression to represent the balance in the savings
account.

Lesson 6-17

Changing the Vertex

NAME _____ DATE _____ PERIOD _____

Learning Goal Let's write new quadratic equations in vertex form to produce certain graphs.

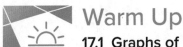

Warm Up
17.1 Graphs of Two Functions

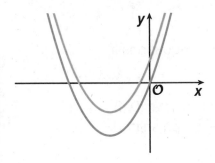

Here are graphs representing the functions f and g, given by $f(x) = x(x + 6)$ and $g(x) = x(x + 6) + 4$.

1. Which graph represents each function? Explain how you know.

2. Where does the graph of f meet the x-axis? Explain how you know.

Activity
17.2 Shifting the Graph

1. How would you change the equation $y = x^2$ so that the vertex of the graph of the new equation is located at the following coordinates and the graph opens as described?

   a. (0, 11), opens up

   b. (7, 11), opens up

   c. (7, -3), opens down

2. Use graphing technology to verify your predictions. Adjust your equations if necessary.

3. Kiran graphed the equation $y = x^2 + 1$ and noticed that the vertex is at (0, 1). He changed the equation to $y = (x - 3)^2 + 1$ and saw that the graph shifted 3 units to the right and the vertex is now at (3, 1).

   Next, he graphed the equation $y = x^2 + 2x + 1$, and observed that the vertex is at (-1, 0). Kiran thought, "If I change the squared term x^2 to $(x - 5)^2$, the graph will move 5 units to the right and the vertex will be at (4, 0)."

   Do you agree with Kiran? Explain or show your reasoning.

NAME _____ DATE _____ PERIOD _____

Activity
17.3 A Peanut Jumping over a Wall

Mai is learning to create computer animation by programming. In one part of her animation, she uses a quadratic function to model the path of the main character, an animated peanut, jumping over a wall.

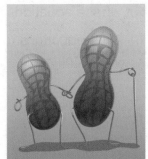

Mai uses the equation $y = -0.1(x - h)^2 + k$ to represent the path of the jump. y represents the height of the peanut as a function of the horizontal distance it travels, x.

On the screen, the base of the wall is located at (22, 0), with the top of the wall at (22, 4.5). The dashed curve in the picture shows the graph of 1 equation Mai tried, where the peanut fails to make it over the wall.

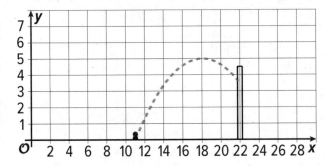

1. What are the values of h and k in this equation?

2. Starting with Mai's equation, choose values for h and k that will guarantee the peanut stays on the screen but also makes it over the wall. Be prepared to explain your reasoning.

Activity

17.4 Smiley Face

Do you see 2 "eyes" and a smiling "mouth" on the graph? The 3 arcs on the graph all represent quadratic functions that were initially defined by $y = x^2$, but whose equations were later modified.

1. Write equations to represent each curve in the smiley face.

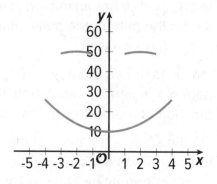

2. What domain is used for each function to create this graph?

NAME _____ DATE _____ PERIOD _____

Summary
Changing the Vertex

The graphs of $y = x^2$, $y = x^2 + 12$ and $y = (x + 3)^2$ all have the same shape but their locations are different. The graph that represents $y = x^2$ has its vertex at $(0, 0)$.

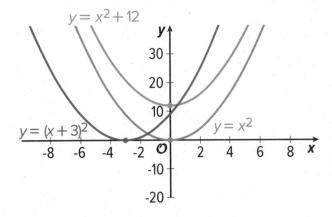

Notice that adding 12 to x^2 raises the graph by 12 units, so the vertex of that graph is at $(0, 12)$. Replacing x^2 with $(x + 3)^2$ shifts the graph 3 units to the left, so the vertex is now at $(-3, 0)$.

We can also shift a graph both horizontally and vertically.

The graph that represents $y = (x + 3)^2 + 12$ will look like that for $y = x^2$ but it will be shifted 12 units up and 3 units to the left. Its vertex is at $(-3, 12)$.

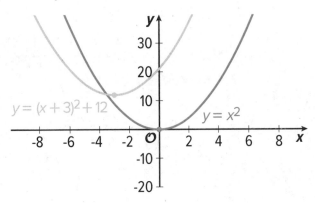

The graph representing the equation $y = -(x + 3)^2 + 12$ has the same vertex at $(-3, 12)$, but because the squared term $(x + 3)^2$ is multiplied by a negative number, the graph is flipped over horizontally, so that it opens downward.

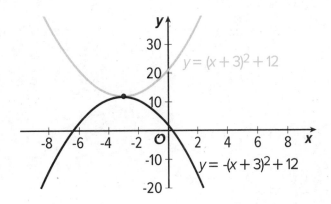

1. Here is the graph of quadratic function f.

   Andre uses the expression $(x - 5)^2 + 7$ to define f.

   Noah uses the expression $(x + 5)^2 - 7$ to define f.

   Do you agree with either of them? Explain your reasoning.

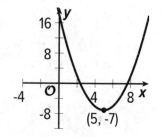

2. Here are the graphs of $y = x^2$, $y = x^2 - 5$, and $y = (x + 2)^2 - 8$.

   a. How do the 3 graphs compare?

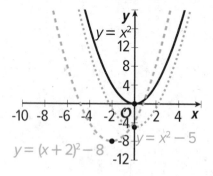

   b. Compare the graphs of $y = x^2$ and $y = x^2 - 5$. What role does the $- 5$ play in the comparison?

   c. Compare the graphs of $y = x^2$ and $y = (x + 2)^2 - 8$. What role does the $+ 2$ play in the comparison?

NAME _____ DATE _____ PERIOD _____

3. Which equation represents the graph of $y = x^2 + 2x - 3$ moved 3 units to the left?

Ⓐ $y = x^2 + 2x - 6$

Ⓑ $y = (x + 3)^2 + 2x - 3$

Ⓒ $y = (x + 3)^2 + 2(x + 3)$

Ⓓ $y = (x + 3)^2 + 2(x + 3) - 3$

4. Select **all** the equations with a graph whose vertex has *both* a positive *x*- and a positive *y*-coordinate.

Ⓐ $y = x^2$

Ⓑ $y = (x - 1)^2$

Ⓒ $y = (x - 3)^2 + 2$

Ⓓ $y = 2(x - 4)^2 - 5$

Ⓔ $y = 0.5(x + 2)^2 + 6$

Ⓕ $y = -(x - 4)^2 + 3$

Ⓖ $y = -2(x - 3)^2 + 1$

5. The height in feet of a soccer ball is modeled by the equation $g(t) = 2 + 50t - 16t^2$, where time *t* is measured in seconds after it was kicked. (Lesson 6-14)

a. How far above the ground was the ball when kicked?

b. What was the initial upward velocity of the ball?

c. Why is the coefficient of the squared term negative?

6. Respond to each question. (Lesson 6-16)

   a. What is the vertex of the graph of the function f defined by $f(x) = -(x - 3)^2 + 6$?

   b. Identify the y-intercept and one other point on the graph of this function.

   c. Sketch the graph of f.

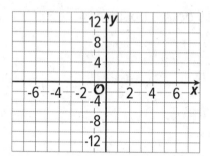

7. At 6:00 a.m., Lin began hiking. At noon, she had hiked 12 miles. At 4:00 p.m., Lin finished hiking with a total trip of 26 miles. (Lesson 4-7)

During which time interval was Lin hiking faster? Explain how you know.

NAME _____ DATE _____ PERIOD _____

8. Kiran bought a slushie every day for a week. Slushies cost $3 each. The amount of money he spends, in dollars, is a function of the number of days of buying slushies. (**Lesson 4-11**)

  a. Sketch a graph of this function. Be sure to label the axes.

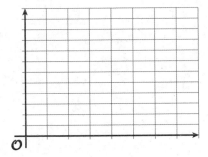

  b. Describe the domain and range of this function.

9. A deposit of $500 has been made in an interest-bearing account. No withdrawals or other deposits (aside from earned interest) are made for 5 years. (**Lesson 5-18**)

  Write an expression to represent the account balance for each of the following situations.

  a. 6.5% annual interest calculated monthly

  b. 6.5% annual interest calculated every two months

  c. 6.5% annual interest calculated quarterly

  d. 6.5% annual interest calculated semi-annually

10. *Technology required.* Function h is defined by $h(x) = 5x + 7$ and function k is defined by $k(x) = (1.005)^x$. (Lesson 5-19)

a. Complete the table with values of $h(x)$ and $k(x)$. When necessary, round to 2 decimal places.

x	h(x)	k(x)
1		
10		
50		
100		

b. Which function do you think *eventually* grows faster? Explain your reasoning.

c. Use graphing technology to verify your answer to the previous question.

Learning Targets

Lesson	Learning Target(s)
6-1 A Different Kind of Change	• I can create drawings, tables, and graphs that represent the area of a garden. • I can recognize a situation represented by a graph that increases then decreases.
6-2 How Does it Change?	• I can describe how a pattern is growing. • I can tell whether a pattern is growing linearly, exponentially, or quadratically. • I know an expression with a squared term is called quadratic.
6-3 Building Quadratic Functions from Geometric Patterns	• I can recognize quadratic functions written in different ways. • I can use information from a pattern of shapes to write a quadratic function. • I know that, in a pattern of shapes, the step number is the input and the number of squares is the output.

(continued on the next page)

(continued from the previous page)

Lesson	Learning Target(s)
6-4 Comparing Quadratic and Exponential Functions	• I can explain using graphs, tables, or calculations that exponential functions eventually grow faster than quadratic functions.
6-5 Building Quadratic Functions to Describe Situations (Part 1)	• I can explain the meaning of the terms in a quadratic expression that represents the height of a falling object. • I can use tables, graphs and equations to represent the height of a falling object.
6-6 Building Quadratic Functions to Describe Situations (Part 2)	• I can create quadratic functions and graphs that represent a situation. • I can relate the vertex of a graph and the zeros of a function to a situation. • I know that the domain of a function can depend on the situation it represents.
6-7 Building Quadratic Functions to Describe Situations (Part 3)	• I can choose a domain that makes sense in a revenue situation. • I can model revenue with quadratic functions and graphs. • I can relate the vertex of a graph and the zeros of a function to a revenue situation.

(continued on the next page)

(continued from the previous page)

Lesson	Learning Target(s)
6-8 Equivalent Quadratic Expressions	• I can rewrite quadratic expressions in different forms by using an area diagram or the distributive property.
6-9 Standard Form and Factored Form	• I can rewrite quadratic expressions given in factored form in standard form using either the distributive property or a diagram. • I know the difference between "factored form" and "standard form."
6-10 Graphs of Functions in Standard and Factored Forms	• I can explain the meaning of the intercepts on a graph of a quadratic function in terms of the situation it represents. • I know how the numbers in the factored form of a quadratic expression relate to the intercepts of its graph.
6-11 Graphing from the Factored Form	• I can graph a quadratic function given in factored form. • I know how to find the vertex and y-intercept of the graph of a quadratic function in factored form without graphing it first.

(continued on the next page)

(continued from the previous page)

Lesson	Learning Target(s)
6-12 Graphing the Standard Form (Part 1)	• I can explain how the a and c in $y = ax^2 + bx + c$ affect the graph of the equation. • I understand how graphs, tables, and equations that represent the same quadratic function are related.
6-13 Graphing the Standard Form (Part 2)	• I can explain how the b in $y = ax^2 + bx + c$ affects the graph of the equation. • I can match equations given in standard and factored form with their graph.
6-14 Graphs That Represent Situations	• I can explain how a quadratic equation and its graph relate to a situation.
6-15 Vertex Form	• I can recognize the "vertex form" of a quadratic equation. • I can relate the numbers in the vertex form of a quadratic equation to its graph.

(continued on the next page)

(continued from the previous page)

Lesson	Learning Target(s)
6-16 Graphing from the Vertex Form	• I can graph a quadratic function given in vertex form, showing a maximum or minimum and the *y*-intercept.
	• I know how to find a maximum or a minimum of a quadratic function given in vertex form without first graphing it.
6-17 Changing the Vertex	• I can describe how changing a number in the vertex form of a quadratic function affects its graph.

Notes:

_____

_____

_____

_____

_____

_____

_____

_____

_____

(continued on the next page)

(continued from the previous page)

Quadratic Equations

Knowing the rules and uses of quadratic equations helps the carpenter complete projects. You'll learn more about quadratic equations in this unit.

Topics

- Finding Unknown Inputs
- Solving Quadratic Equations
- Completing the Square
- The Quadratic Formula
- Vertex Form Revisited
- Putting It All Together

Unit 7

Quadratic Equations

Completing the Square

The Quadratic Formula

Vertex Form Revisited

Putting It All Together

Lesson 7-1

Finding Unknown Inputs

NAME _____ DATE _____ PERIOD _____

Learning Goal Let's find some new equations to solve.

 ## Warm Up
1.1 What Goes Up Must Come Down

A mechanical device is used to launch a potato vertically into the air. The potato is launched from a platform 20 feet above the ground, with an initial vertical velocity of 92 feet per second.

The function $h(t) = -16t^2 + 92t + 20$ models the height of the potato over the ground, in feet, t seconds after launch.

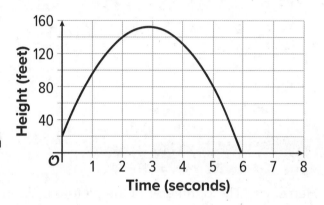

Here is the graph representing the function.

For each question, be prepared to explain your reasoning.

1. What is the height of the potato 1 second after launch?

2. 8 seconds after launch, will the potato still be in the air?

3. Will the potato reach 120 feet? If so, when will it happen?

4. When will the potato hit the ground?

Activity
1.2 A Trip to the Frame Shop

Your teacher will give you a picture that is 7 inches by 4 inches, a piece of framing material measuring 4 inches by 2.5 inches, and a pair of scissors.

Cut the framing material to create a rectangular frame for the picture. The frame should have the same thickness all the way around and have no overlaps. All of the framing material should be used (with no leftover pieces). Framing material is very expensive! You get 3 copies of the framing material, in case you make mistakes and need to recut.

Are you ready for more?

Han says, "The perimeter of the picture is 22 inches. If I cut the framing material into 9 pieces, each one being 2.5 inches by $\frac{4}{9}$ inch, I'll have more than enough material to surround the picture because those pieces would mean 22.5 inches for the frame."

Do you agree with Han? Explain your reasoning.

NAME _____ DATE _____ PERIOD _____

 Activity

1.3 Representing the Framing Problem

Here is a diagram that shows the picture with a frame that is the same thickness all the way around. The picture is 7 inches by 4 inches. The frame is created from 10 square inches of framing material (in the form of a rectangle measuring 4 inches by 2.5 inches).

1. Write an equation to represent the relationship between the measurements of the picture and of the frame, and the area of the framed picture. Be prepared to explain what each part of your equation represents.

2. What would a solution to this equation mean in this situation?

 Summary

Finding Unknown Inputs

The height of a softball, in feet, t seconds after someone throws it straight up, can be defined by $f(t) = -16t^2 + 32t + 5$. The input of function f is time, and the output is height.

We can find the output of this function at any given input. For instance:

- At the beginning of the softball's journey, when $t = 0$, its height is given by $f(0)$.

- Two seconds later, when $t = 2$, its height is given by $f(2)$.

The values of $f(0)$ and $f(2)$ can be found using a graph or by evaluating the expression $-16t^2 + 32t + 5$ at those values of t.

What if we know the output of the function and want to find the inputs? For example:

- When does the softball hit the ground?

   Answering this question means finding the values of t that make $f(t) = 0$, or solving $-16t^2 + 32t + 5 = 0$.

- How long will it take the ball to reach 8 feet?

   This means finding one or more values of t that make $f(t) = 8$, or solving the equation $-16t^2 + 32t + 5 = 8$.

The equations $-16t^2 + 32t + 5 = 0$ and $-16t^2 + 32t + 5 = 8$ are *quadratic equations*. One way to solve these equations is by graphing $y = f(t)$.

- To answer the first question, we can look for the horizontal intercepts of the graph, where the vertical coordinate is 0.

- To answer the second question, we can look for the horizontal coordinates that correspond to a vertical coordinate of 8.

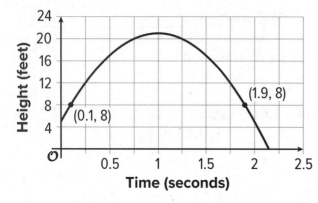

We can see that there are two solutions to the equation $-16t^2 + 32t + 5 = 8$.

The softball has a height of 8 feet twice, when going up and when coming down, and these occur when t is about 0.1 or 1.9.

Often, when we are modeling a situation mathematically, an approximate solution is good enough. Sometimes, however, we would like to know exact solutions, and it may not be possible to find them using a graph.

In this unit, we will learn more about quadratic equations and how to solve them exactly using algebraic techniques.

Glossary

quadratic expression

NAME _____ DATE _____ PERIOD _____

Practice
Finding Unknown Inputs

1. A girl throws a paper airplane from her treehouse. The height of the plane is a function of time and can be modeled by the equation $h(t) = 25 + 2.5t - \frac{1}{2}t^2$. Height is measured in feet and time is measured in seconds after the airplane is thrown.

   a. Evaluate $h(0)$ and explain what this value means in this situation.

   b. What would a solution to $h(t) = 0$ mean in this situation?

   c. What does the equation $h(9) = 7$ mean?

   d. What does the model say about the airplane 2.5 seconds after the girl throws it if each of these statements is true?

   $h(2) = 28$ $h(2.5) = 28.125$ $h(3) = 28$

2. A square picture has a frame that is 3 inches thick all the way around. The total side length of the picture and frame is x inches.

   Which expression represents the area of the square picture, without the frame? If you get stuck, try sketching a diagram.

   A. $(2x + 3)(2x + 3)$

   B. $(x + 6)(x + 6)$

   C. $(2x - 3)(2x - 3)$

   D. $(x - 6)(x - 6)$

3. The revenue from a youth league baseball game depends on the price per ticket, x.

Here is a graph that represents the revenue function, R.

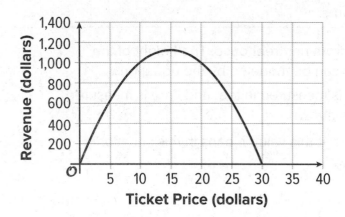

Select **all** the true statements.

(A.) $R(5)$ is a little more than 600.

(B.) $R(600)$ is a little less than 5.

(C.) The maximum possible ticket price is $15.

(D.) The maximum possible revenue is about $1,125.

(E.) If tickets cost $10, the predicted revenue is $1,000.

(F.) If tickets cost $20, the predicted revenue is $1,000.

NAME _____ DATE _____ PERIOD _____

4. A garden designer designed a square decorative pool. The pool is surrounded by a walkway.

   On two opposite sides of the pool, the walkway is 8 feet. On the other two opposite sides, the walkway is 10 feet.

   Here is a diagram of the design.

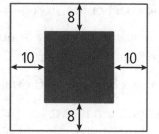

   The final design for the pool and walkway covers a total area of 1,440 square feet.

   a. The side length of the square pool is x. Write an expression that represents:

      i. the total length of the rectangle (including the pool and walkway)

      ii. the total width of the rectangle (including the pool and walkway)

      iii. the total area of the pool and walkway

   b. Write an equation of the form: Your expression = 1,440. What does a solution to the equation mean in this situation?

5. Suppose m and c each represent the position number of a letter in the alphabet, but m represents the letters in the original message and c represents the letters in a secret code. The equation $c = m + 2$ is used to encode a message. (Lesson 4-15)

   a. Write an equation that can be used to decode the secret code into the original message.

   b. What does this code say: "OCVJ KU HWP!"?

6. An American traveler who is heading to Europe is exchanging some U.S. dollars for European euros. At the time of his travel, 1 dollar can be exchanged for 0.91 euros. (Lesson 4-15)

   a. Find the amount of money in euros that the American traveler would get if he exchanged 100 dollars.

   b. What if he exchanged 500 dollars?

   c. Write an equation that gives the amount of money in euros, e, as a function of the dollar amount being exchanged, d.

   d. Upon returning to America, the traveler has 42 euros to exchange back into U.S. dollars. How many dollars would he get if the exchange rate is still the same?

   e. Write an equation that gives the amount of money in dollars, d, as a function of the euro amount being exchanged, e.

7. A random sample of people are asked to give a taste score—either "low" or "high"—to two different types of ice cream. The two types of ice cream have identical formulas, except they differ in the percentage of sugar in the ice cream.

   What values could be used to complete the table so that it suggests there is an association between taste score and percentage of sugar? Explain your reasoning. (Lesson 3-3)

	12% Sugar	15% Sugar
low taste score	239	
high taste score	126	

Lesson 7-2

When and Why Do We Write Quadratic Equations?

NAME _____ DATE _____ PERIOD _____

Learning Goal Let's try to solve some quadratic equations.

Warm Up
2.1 How Many Tickets?

The expression $12t + 2.50$ represents the cost to purchase tickets for a play, where t is the number of tickets. Be prepared to explain your response to each question.

1. A family paid \$62.50 for tickets. How many tickets were bought?

2. A teacher paid \$278.50 for tickets for her students. How many tickets were bought?

Activity
2.2 The Flying Potato Again

The other day, you saw an equation that defines the height of a potato as a function of time after it was launched from a mechanical device. Here is a different function modeling the height of a potato, in feet, t seconds after being fired from a different device:

$$f(t) = -16t^2 + 80t + 64$$

1. What equation would we solve to find the time at which the potato hits the ground?

2. Use any method *except graphing* to find a solution to this equation.

Activity

2.3 Revenue from Ticket Sales

The expressions $p(200 - 5p)$ and $-5p^2 + 200p$ define the same function. The function models the revenue a school would earn from selling raffle tickets at p dollars each.

1. At what price or prices would the school collect $0 revenue from raffle sales? Explain or show your reasoning.

2. The school staff noticed that there are two ticket prices that would both result in a revenue of $500. How would you find out what those two prices are?

Are you ready for more?

Can you find the following prices without graphing?

1. If the school charges $10, it will collect $1,500 in revenue. Find another price that would generate $1,500 in revenue.

2. If the school charges $28, it will collect $1,680 in revenue. Find another price that would generate $1,680 in revenue.

3. Find the price that would produce the maximum possible revenue. Explain your reasoning.

NAME _____ DATE _____ PERIOD _____

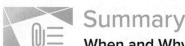

Summary
When and Why Do We Write Quadratic Equations?

The height of a potato that is launched from a mechanical device can be modeled by a function, g with x representing time in seconds. Here are two expressions that are equivalent and both define function g.

$$-16x^2 + 80x + 96 \qquad\qquad -16(x - 6)(x + 1)$$

Notice that one expression is in *standard form* and the other is in *factored form*.

Suppose we wish to know, without graphing the function, the time when the potato will hit the ground. We know that the value of the function at that time is 0, so we can write:

$$-16x^2 + 80x + 96 = 0 \qquad\qquad -16(x - 6)(x + 1) = 0$$

Let's try solving $-16x^2 + 80x + 96 = 0$, using some familiar moves. For example:

- Subtract 96 from each side: $-16x^2 + 80x = -96$

- Apply the distributive property to rewrite the expression on the left: $-16(x^2 - 5x) = -96$

- Divide both sides by -16: $x^2 - 5x = 6$

- Apply the distributive property to rewrite the expression on the left: $x(x - 5) = 6$

These steps don't seem to get us any closer to a solution. We need some new moves!

What if we use the other equation? Can we find the solutions to $-16(x - 6)(x + 1) = 0$?

Earlier, we learned that the *zeros* of a quadratic function can be identified when the expression defining the function is in factored form. The solutions to $-16(x - 6)(x + 1) = 0$ are the zeros to function g, so this form may be more helpful! We can reason that:

- If x is 6, then the value of $x - 6$ is 0, so the entire expression has a value of 0.

- If x is -1, then the value of $x + 1$ is 0, so the entire expression also has a value of 0.

This tells us that 6 and -1 are solutions to the equation, and that the potato hits the ground after 6 seconds. (A negative value of time is not meaningful, so we can disregard the -1.)

Both equations we see here are **quadratic equations**. In general, a quadratic equation is an equation that can be expressed as $ax^2 + bx + c = 0$, where a, b, and c are constants and $a \neq 0$.

In upcoming lessons, we will learn how to rewrite quadratic equations into forms that make the solutions easy to see.

Glossary

factored form (of a quadratic expression)

quadratic equation

standard form (of a quadratic expression)

zero (of a function)

NAME _____ DATE _____ PERIOD _____

Practice
When and Why Do We Write Quadratic Equations?

1. A set of kitchen containers can be stacked to save space. The height of the stack is given by the expression $1.5c + 7.6$, where c is the number of containers.

   a. Find the height of a stack made of 8 containers.

   b. A tower made of all the containers is 40.6 cm tall. How many containers are in the set?

   c. Noah looks at the equation and says, "7.6 must be the height of a single container." Do you agree with Noah? Explain your reasoning.

2. Select **all** values of x that are solutions to the equation $(x - 5)(7x - 21) = 0$.

   (A.) -7   (D.) 0   (F.) 5

   (B.) -5   (E.) 3   (G.) 7

   (C.) -3

3. The expressions $30x^2 - 105x - 60$ and $(5x - 20)(6x + 3)$ define the same function, f.

   a. Which expression makes it easier to find $f(0)$? Explain your reasoning.

   b. Find $f(0)$.

c. Which expression makes it easier to find the values of x that make the equation $f(x) = 0$ true? Explain or show your reasoning.

d. Find the values of x that make $f(x) = 0$.

4. A band is traveling to a new city to perform a concert. The revenue from their ticket sales is a function of the ticket price, x, and can be modeled with $(x - 6)(250 - 5x)$.

   What are the ticket prices at which the band would make no money at all?

5. Two students built a small rocket from a kit and attached an altimeter (a device for recording altitude or height) to the rocket. They record the height of the rocket over time since it is launched in the table, based on the data from the altimeter.

Time (Seconds)	0	1	3	4	7	8
Height (Meters)	0	110.25	236.25	252	110.25	0

Function h gives the height in meters as a function of time in seconds, t.
(Lesson 7-1)

   a. What is the value of $h(3)$?

   b. What value of t gives $h(t) = 252$?

NAME _____ DATE _____ PERIOD _____

c. Explain why $h(0) = h(8)$.

d. Based on the data, which equation about the function could be true: $h(2) = 189$ or $h(189) = 2$? Explain your reasoning.

6. The screen of a tablet has dimensions 8 inches by 5 inches. The border around the screen has thickness x. **(Lesson 7-1)**

   a. Write an expression for the total area of the tablet, including the frame.

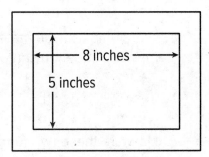

   b. Write an equation for which your expression is equal to 50.3125. Explain what a solution to this equation means in this situation.

   c. Try to find the solution to the equation. If you get stuck, try guessing and checking. It may help to think about tablets that you have seen.

7. Here are a few pairs of positive numbers whose sum is 15. The pair of numbers that have a sum of 15 and will produce the largest possible product is *not* shown.

   Find this pair of numbers.

First Number	Second Number	Product
1	14	14
3	12	36
5	10	50
7	8	56

8. Kilometer is a measurement in the metric system, while mile is a measurement in the customary system. One kilometer equals approximately 0.621 mile. **(Lesson 4-15)**

   a. The number of miles, m, is a function of the number of kilometers, k. What equation can be written to represent this function?

   b. The number of kilometers, k, is a function of the number of miles, m. What equation can be written to represent this function?

   c. How are these two functions related? Explain how you know.

Lesson 7-3

Solving Quadratic Equations by Reasoning

NAME _____ DATE _____ PERIOD _____

Learning Goal Let's find solutions to quadratic equations.

Warm Up
3.1 How Many Solutions?

How many solutions does each equation have? What are the solution(s)? Be prepared to explain how you know.

1. $x^2 = 9$

2. $x^2 = 0$

3. $x^2 - 1 = 3$

4. $2x^2 = 50$

5. $(x + 1)(x + 1) = 0$

6. $x(x - 6) = 0$

7. $(x - 1)(x - 1) = 4$

Activity
3.2 Finding Pairs of Solutions

Each of these equations has two solutions. What are they? Explain or show your reasoning.

1. $n^2 + 4 = 404$

2. $432 = 3n^2$

3. $144 = (n + 1)^2$

4. $(n - 5)^2 - 30 = 70$

Are you ready for more?

1. How many solutions does the equation $(x - 3)(x + 1)(x + 5) = 0$ have? What are the solutions?

2. How many solutions does the equation $(x - 2)(x - 7)(x - 2) = 0$ have? What are the solutions?

3. Write a new equation that has 10 solutions.

NAME _____ DATE _____ PERIOD _____

Summary
Solving Quadratic Equations by Reasoning

Some quadratic equations can be solved by performing the same operation to each side of the equal sign and reasoning about values of the variable would make the equation true.

Suppose we wanted to solve $3(x + 1)^2 - 75 = 0$. We can proceed like this:

- Add 75 to each side: $\qquad\qquad\qquad\qquad\qquad 3(x + 1)^2 = 75$
- Divide each side by 3: $\qquad\qquad\qquad\qquad\quad (x + 1)^2 = 25$
- What number can be squared to get 25? $\qquad\quad (\;\square\;)^2 = 25$
- There are two numbers that work, 5 and -5: $\quad 5^2 = 25$ and $(-5)^2 = 25$
- If $x + 1 = 5$, then $x = 4$.
- If $x + 1 = -5$, then $x = -6$.

This means that both $x = 4$ and $x = -6$ make the equation true and are solutions to the equation.

1. Consider the equation $x^2 = 9$.

   a. Show that 3, -3, $\sqrt{9}$, and -$\sqrt{9}$ are each a solution to the equation.

   b. Show that 9 and $\sqrt{3}$ are each *not* a solution to the equation.

2. Solve $(x - 1)^2 = 16$. Explain or show your reasoning.

3. Here is one way to solve the equation $\frac{5}{9}y^2 = 5$. Explain what is done in each step.

$$\frac{5}{9}y^2 = 5 \qquad \text{Original equation}$$

$$5y^2 = 45 \qquad \text{Step 1}$$

$$y^2 = 9 \qquad \text{Step 2}$$

$$y = 3 \quad \text{or} \quad y = -3 \qquad \text{Step 3}$$

NAME _____ DATE _____ PERIOD _____

4. Diego and Jada are working together to solve the quadratic equation $(x - 2)^2 = 100$.

Diego solves the equation by dividing each side of the equation by 2 and then adding 2 to each side. He writes:

$(x - 2) = 50$
$x = 52$

Jada asks Diego why he divides each side by 2 and he says, "I want to find a number that equals 100 when multiplied by itself. That number is half of 100."

a. What mistake is Diego making?

b. If you were Jada, what could you say to Diego to help him realize his mistake?

5. As part of a publicity stunt (an event designed to draw attention), a TV host drops a watermelon from the top of a tall building. The height of the watermelon t seconds after it is dropped is given by the function $h(t) = 850 - 16t^2$, where h is in feet. **(Lesson 7-1)**

a. Find $h(4)$. Explain what this value means in this situation.

b. Find $h(0)$. What does this value tell us about the situation?

c. Is the watermelon still in the air 8 seconds after it is dropped? Explain how you know.

6. A zoo offers unlimited drink refills to visitors who purchase its souvenir cup. The cup and the first fill cost $10, and refills after that are $2 each. The expression $10 + 2r$ represents the total cost of the cup and r refills. (Lesson 7-2)

a. A family visited the zoo several times over a summer. That summer, they paid $30 for one cup and multiple refills. How many refills did they buy?

b. A visitor has $18 to spend on drinks at the zoo today and buys a souvenir cup. How many refills can they afford during the visit?

c. Another visitor spent $10 on this deal. Did they buy any refills? Explain how you know.

7. Here are a few pairs of positive numbers whose sum is 15. The pair of numbers that have a sum of 15 and will produce the largest possible product is *not* shown. (Lesson 6-1)

Find this pair of numbers.

First Number	Second Number	Product
1	14	14
3	12	36
5	10	50
7	8	56

NAME _____ DATE _____ PERIOD _____

8. Clare is 5 years older than her sister. **(Lesson 4-15)**

    a. Write an equation that defines her sister's age, s, as a function of Clare's age, c.

    b. Write an equation that defines Clare's age, c, as a function of her sister's age, s.

    c. Graph each function. Be sure to label the axes.

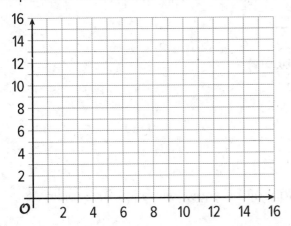

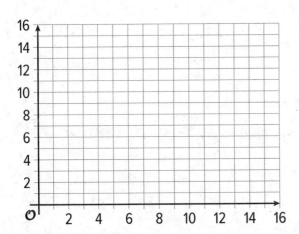

    d. Describe how the two graphs compare.

9. The graph shows the weight of snow as it melts. The weight decreases exponentially. (Lesson 5-5)

a. By what factor does the weight of the snow decrease each hour? Explain how you know.

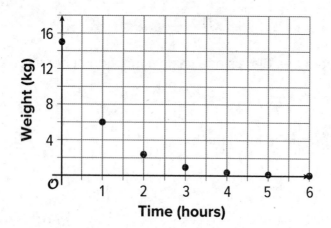

b. Does the graph predict that the weight of the snow will reach 0? Explain your reasoning.

c. Will the weight of the actual snow, represented by the graph, reach 0? Explain how you know.

Lesson 7-4

Solving Quadratic Equations with the Zero Product Property

NAME _____ DATE _____ PERIOD _____

Learning Goal Let's find solutions to equations that contain products that equal zero.

Warm Up
4.1 Math Talk: Solve These Equations

What values of the variables make each equation true?

$6 + 2a = 0$

$7b = 0$

$7(c - 5) = 0$

$g \cdot h = 0$

Activity
4.2 Take the Zero Product Property Out for a Spin

For each equation, find its solution or solutions. Be prepared to explain your reasoning.

1. $x - 3 = 0$

2. $x + 11 = 0$

3. $2x + 11 = 0$

4. $x(2x + 11) = 0$

5. $(x - 3)(x + 11) = 0$

6. $(x - 3)(2x + 11) = 0$

7. $x(x + 3)(3x - 4) = 0$

1. Use factors of 48 to find as many solutions as you can to the equation $(x - 3)(x + 5) = 48$.

2. Once you think you have all the solutions, explain why these must be the only solutions.

Activity

4.3 Revisiting a Projectile

We have seen quadratic functions modeling the height of a projectile as a function of time.

Here are two ways to define the same function that approximates the height of a projectile in meters, t seconds after launch:

$$h(t) = -5t^2 + 27t + 18 \qquad\qquad h(t) = (-5t - 3)(t - 6)$$

1. Which way of defining the function allows us to use the zero product property to find out when the height of the object is 0 meters?

2. Without graphing, determine at what time the height of the object is 0 meters. Show your reasoning.

NAME _____ DATE _____ PERIOD _____

Summary
Solving Quadratic Equations with the Zero Product Property

The **zero product property** says that if the product of two numbers is 0, then one of the numbers must be 0. In other words, if $a \cdot b = 0$, then either $a = 0$ or $b = 0$. This property is handy when an equation we want to solve states that the product of two factors is 0.

Suppose we want to solve $m(m + 9) = 0$. This equation says that the product of m and $(m + 9)$ is 0. For this to be true, either $m = 0$ or $m + 9 = 0$, so both 0 and -9 are solutions.

Here is another equation: $(u - 2.345)(14u + 2) = 0$. The equation says the product of $(u - 2.345)$ and $(14u + 2)$ is 0, so we can use the zero product property to help us find the values of u. For the equation to be true, one of the factors must be 0.

- For $u - 2.345 = 0$ to be true, u would have to be 2.345.
- For $14u + 2 = 0$ or $14u = -2$ to be true, u would have to be $-\frac{2}{14}$ or $-\frac{1}{7}$.

The solutions are 2.345 and $-\frac{1}{7}$.

In general, when a quadratic expression in factored form is on one side of an equation and 0 is on the other side, we can use the zero product property to find its solutions.

Glossary

zero product property

Practice
Solving Quadratic Equations with the Zero Product Property

1. If the equation $(x + 10)x = 0$ is true, which statement is also true according to the zero product property?

   (A.) only $x = 0$

   (B.) either $x = 0$ or $x + 10 = 0$

   (C.) either $x^2 = 0$ or $10x = 0$

   (D.) only $x + 10 = 0$

2. What are the solutions to the equation $(10 - x)(3x - 9) = 0$?

   (A.) -10 and 3

   (B.) -10 and 9

   (C.) 10 and 3

   (D.) 10 and 9

3. Solve each equation.

   a. $(x - 6)(x + 5) = 0$

   b. $(x - 3)\left(\frac{2}{3}x - 6\right) = 0$

   c. $(-3x - 15)(x + 7) = 0$

4. Consider the expressions $(x - 4)(3x - 6)$ and $3x^2 - 18x + 24$.

   Show that the two expressions define the same function.

5. Kiran saw that if the equation $(x + 2)(x - 4) = 0$ is true, then, by the zero product property, either $x + 2$ is 0 or $x - 4$ is 0. He then reasoned that, if $(x + 2)(x - 4) = 72$ is true, then either $x + 2$ is equal to 72 or $x - 4$ is equal to 72. Explain why Kiran's conclusion is incorrect.

NAME _____ DATE _____ PERIOD _____

6. Andre wants to solve the equation $5x^2 - 4x - 18 = 20$. He uses a graphing calculator to graph $y = 5x^2 - 4x - 18$ and $y = 20$ and finds that the graphs cross at the points (-2.39,20) and (3.19,20). **(Lesson 7-2)**

   a. Substitute each x-value Andre found into the expression $5x^2 - 4x - 18$. Then evaluate the expression.

   b. Why did neither solution make $5x^2 - 4x - 18$ equal exactly 20?

7. Select **all** the solutions to the equation $7x^2 = 343$. **(Lesson 7-3)**

   (A.) 49   (C.) 7

   (B.) $-\sqrt{7}$ (D.) -7

   (E.) $\sqrt{49}$ (F.) $\sqrt{-49}$

   (G.) $-\sqrt{49}$

8. Here are two graphs that correspond to two patients, A and B. Each graph shows the amount of insulin, in micrograms (mcg) in a patient's body h hours after receiving an injection. The amount of insulin in each patient decreases exponentially.

Patient A

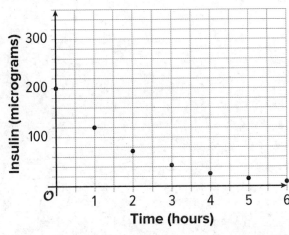

Patient B

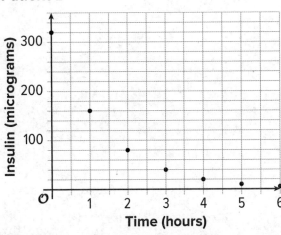

Select **all** statements that are true about the insulin level of the two patients. (Lesson 5-6)

A. After the injection, the patients have the same amount of insulin in their bodies.

B. An equation for the micrograms of insulin, a, in Patient A's body h hours after the injection is $a = 200 \cdot \left(\dfrac{3}{5}\right)^{h}$.

C. The insulin in Patient A is decaying at a faster rate than in Patient B.

D. After 3 hours, Patient A has more insulin in their body than Patient B.

E. At some time between 2 and 3 hours, the patients have the same insulin level.

9. Han says this pattern of dots can be represented by a quadratic relationship because the dots are arranged in a rectangle in each step. (Lesson 6-2)

Do you agree? Explain your reasoning.

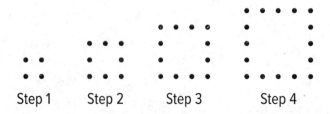

Step 1   Step 2   Step 3   Step 4

Lesson 7-5

How Many Solutions?

NAME _____ DATE _____ PERIOD _____

Learning Goal Let's use graphs to investigate quadratic equations that have two solutions, one solution, or no solutions.

Warm Up
5.1 Math Talk: Four Equations

Decide whether each statement is true or false.

3 is the only solution to $x^2 - 9 = 0$.

A solution to $x^2 + 25 = 0$ is -5.

$x(x - 7) = 0$ has two solutions.

5 and -7 are the solutions to $(x - 5)(x + 7) = 12$.

Han is solving three equations by graphing.

$$(x - 5)(x - 3) = 0$$

$$(x - 5)(x - 3) = -1$$

$$(x - 5)(x - 3) = -4$$

1. To solve the first equation, $(x - 5)(x - 3) = 0$, he graphed $y = (x - 5)(x - 3)$ and then looked for the x-intercepts of the graph.

   a. Explain why the x-intercepts can be used to solve $(x - 5)(x - 3) = 0$.

   b. What are the solutions?

2. To solve the second equation, Han rewrote it as $(x - 5)(x - 3) + 1 = 0$. He then graphed $y = (x - 5)(x - 3) + 1$.

   Use graphing technology to graph $y = (x - 5)(x - 3) + 1$. Then, use the graph to solve the equation. Be prepared to explain how you use the graph for solving.

NAME _____ DATE _____ PERIOD _____

3. Solve the third equation using Han's strategy.

4. Think about the strategy you used and the solutions you found.

   a. Why might it be helpful to rearrange each equation to equal 0 on one side and then graph the expression on the non-zero side?

   b. How many solutions does each of the three equations have?

The equations $(x - 3)(x - 5) = -1$, $(x - 3)(x - 5) = 0$, and $(x - 3)(x - 5) = 3$ all have whole-number solutions.

1. Use graphing technology to graph each of the following pairs of equations on the same coordinate plane. Analyze the graphs and explain how each pair helps to solve the related equation.

   - $y = (x - 3)(x - 5)$ and $y = -1$
   - $y = (x - 3)(x - 5)$ and $y = 0$
   - $y = (x - 3)(x - 5)$ and $y = 3$

2. Use the graphs to help you find a few other equations of the form $(x - 3)(x - 5) = z$ that have whole-number solutions.

NAME _____ DATE _____ PERIOD _____

3. Find a pattern in the values of z that give whole-number solutions.

4. Without solving, determine if $(x - 5)(x - 3) = 120$ and $(x - 5)(x - 3) = 399$ have whole-number solutions. Explain your reasoning.

Activity

5.3 Finding All the Solutions

Solve each equation. Be prepared to explain or show your reasoning.

1. $x^2 = 121$

2. $x^2 - 31 = 5$

3. $(x - 4)(x - 4) = 0$

4. $(x + 3)(x - 1) = 5$

5. $(x + 1)^2 = -4$

6. $(x - 4)(x - 1) = 990$

Activity

5.4 Analyzing Errors in Equation Solving

1. Consider $(x - 5)(x + 1) = 7$. Priya reasons that if this is true, then either $x - 5 = 7$ or $x + 1 = 7$. So, the solutions to the original equation are 12 and 6.

   Do you agree? If not, where was the mistake in Priya's reasoning?

2. Consider $x^2 - 10x = 0$. Diego says to solve we can just divide each side by x to get $x - 10 = 0$, so the solution is 10. Mai says, "I wrote the expression on the left in factored form, which gives $x(x - 10) = 0$, and ended up with two solutions: 0 and 10."

   Do you agree with either strategy? Explain your reasoning.

NAME _____ DATE _____ PERIOD _____

Summary
How Many Solutions?

Quadratic equations can have two, one, or no solutions.

We can find out how many solutions a quadratic equation has and what the solutions are by rearranging the equation into the form of expression $= 0$, graphing the function that the expression defines, and determining its zeros. Here are some examples.

- $x^2 = 5x$

  Let's first subtract $5x$ from each side and rewrite the equation as $x^2 - 5x = 0$. We can think of solving this equation as finding the zeros of a function defined by $x^2 - 5x$.

  If the output of this function is y, we can graph $y = x^2 - 5x$ and identify where the graph intersects the x-axis, where the y-coordinate is 0.

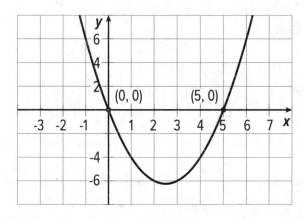

From the graph, we can see that the x-intercepts are $(0, 0)$ and $(5, 0)$, so $x^2 - 5x$ equals 0 when x is 0 and when x is 5.

The graph readily shows that there are two solutions to the equation.

Note that the equation $x^2 = 5x$ can be solved without graphing, but we need to be careful *not* to divide both sides by x. Doing so will give us $x = 5$ but will show no trace of the other solution, $x = 0$!

Even though dividing both sides by the same value is usually acceptable for solving equations, we avoid dividing by the same variable because it may eliminate a solution.

- $(x - 6)(x - 4) = -1$

  Let's rewrite the equation as $(x - 6)(x - 4) + 1 = 0$, and consider it to represent a function defined by $(x - 6)(x - 4) + 1$ and whose output, y, is 0.

  Let's graph $y = (x - 6)(x - 4) + 1$ and identify the x-intercepts.

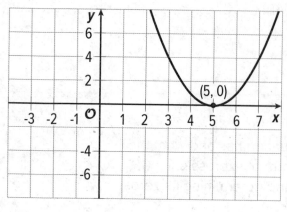

The graph shows one *x*-intercept at (5, 0). This tells us that the function defined by $(x - 6)(x - 4) + 1$ has only one zero.

It also means that the equation $(x - 6)(x - 4) + 1 = 0$ is true only when $x = 5$. The value 5 is the only solution to the equation.

- $(x - 3)(x - 3) = -4$

    Rearranging the equation gives $(x - 3)(x - 3) + 4 = 0$.

    Let's graph $y = (x - 3)(x - 3) + 4$ and find the *x*-intercepts.

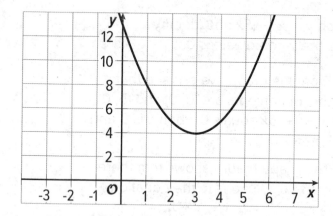

The graph does not intersect the *x*-axis, so there are no *x*-intercepts.

This means there are no *x*-values that can make the expression $(x - 3)(x - 3) + 4$ equal 0, so the function defined by $y = (x - 3)(x - 3) + 4$ has no zeros.

The equation $(x - 3)(x - 3) = -4$ has no solutions.

We can see that this is the case even without graphing. $(x - 3)(x - 3) = -4$ is $(x - 3)^2 = -4$. Because no number can be squared to get a negative value, the equation has no solutions.

Earlier you learned that graphing is not always reliable for showing precise solutions. This is still true here. The *x*-intercepts of a graph are not always whole-number values. While they can give us an idea of how many solutions there are and what the values may be (at least approximately), for exact solutions we still need to rely on algebraic ways of solving.

NAME _____ DATE _____ PERIOD _____

Practice
How Many Solutions?

1. Rewrite each equation so that the expression on one side could be graphed and the x-intercepts of the graph would show the solutions to the equation.

   a. $3x^2 = 81$

   b. $(x - 1)(x + 1) - 9 = 5x$

   c. $x^2 - 9x + 10 = 32$

   d. $6x(x - 8) = 29$

2. Respond to each queation.

   a. Here are equations that define quadratic functions f, g, and h. Sketch a graph, by hand or using technology, that represents each equation.

   $f(x) = x^2 + 4$ $g(x) = x(x + 3)$ $h(x) = (x - 1)^2$

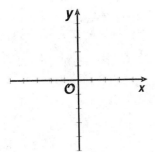   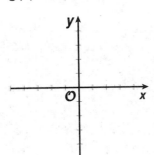   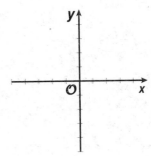

   b. Determine how many solutions each $f(x) = 0$, $g(x) = 0$, and $h(x) = 0$ has. Explain how you know.

3. Mai is solving the equation $(x - 5)^2 = 0$. She writes that the solutions are $x = 5$ and $x = -5$. Han looks at her work and disagrees. He says that only $x = 5$ is a solution. Who do you agree with? Explain your reasoning.

4. The graph shows the number of square meters, A, covered by algae in a lake w weeks after it was first measured.

In a second lake, the number of square meters, B, covered by algae is defined by the equation $B = 975 \cdot \left(\frac{2}{5}\right)^w$, where w is the number of weeks since it was first measured.

For which algae population is the area decreasing more rapidly? Explain how you know. (Lesson 5-6)

5. If the equation $(x - 4)(x + 6) = 0$ is true, which is also true according to the zero product property? (Lesson 7-4)

(A.) only $x - 4 = 0$

(B.) only $x + 6 = 0$

(C.) $x - 4 = 0$ or $x + 6 = 0$

(D.) $x = -4$ or $x = 6$

NAME _____ DATE _____ PERIOD _____

6. Respond to each question. **(Lesson 7-3)**

    a. Solve the equation $25 = 4z^2$.

    b. Show that your solution or solutions are correct.

7. To solve the quadratic equation $3(x - 4)^2 = 27$, Andre and Clare wrote the following:

Andre   Clare

$$3(x - 4)^2 = 27 \qquad\qquad 3(x - 4)^2 = 27$$
$$(x - 4)^2 = 9 \qquad\qquad (x - 4)^2 = 9$$
$$x^2 - 4^2 = 9 \qquad\qquad x - 4 = 3$$
$$x^2 - 16 = 9 \qquad\qquad x = 7$$
$$x^2 = 25$$
$$x = 5 \quad \text{or} \quad x = \text{-}5$$

(Lesson 7-3)

    a. Identify the mistake each student made.

    b. Solve the equation and show your reasoning.

8. Decide if each equation has 0, 1, or 2 solutions and explain how you know.

a. $x^2 - 144 = 0$

b. $x^2 + 144 = 0$

c. $x(x - 5) = 0$

d. $(x - 8)^2 = 0$

e. $(x + 3)(x + 7) = 0$

Lesson 7-6

Rewriting Quadratic Expressions in Factored Form (Part 1)

NAME _____ DATE _____ PERIOD _____

Learning Goal Let's write expressions in factored form.

Warm Up
6.1 Puzzles of Rectangles

Here are two puzzles that involve side lengths and areas of rectangles.
Can you find the missing area in Figure A and the missing length in Figure B?
Be prepared to explain your reasoning.

Figure A

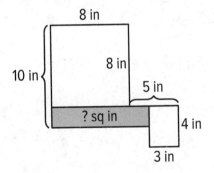

Figure B

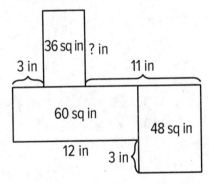

Activity

6.2 Using Diagrams to Understand Equivalent Expressions

1. Use a diagram to show that each pair of expressions is equivalent.

   $x(x + 3)$ and $x^2 + 3x$ $x(x + \text{-}6)$ and $x^2 - 6x$

   $(x + 2)(x + 4)$ and $x^2 + 6x + 8$ $(x + 4)(x + 10)$ and $x^2 + 14x + 40$

   $(x + \text{-}5)(x + \text{-}1)$ and $x^2 - 6x + 5$ $(x - 1)(x - 7)$ and $x^2 - 8x + 7$

2. Observe the pairs of expressions that involve the product of two sums or two differences. How is each expression in factored form related to the equivalent expression in standard form?

NAME _____ DATE _____ PERIOD _____

Activity
6.3 Let's Rewrite Some Expressions!

Each row in the table contains a pair of equivalent expressions.

Complete the table with the missing expressions. If you get stuck, consider drawing a diagram.

Factored Form	Standard Form
$x(x + 7)$	
	$x^2 + 9x$
	$x^2 - 8x$
$(x + 6)(x + 2)$	
	$x^2 + 13x + 12$
$(x - 6)(x - 2)$	
	$x^2 - 7x + 12$
	$x^2 + 6x + 9$
	$x^2 + 10x + 9$
	$x^2 - 10x + 9$
	$x^2 - 6x + 9$
	$x^2 + (m + n)x + mn$

A mathematician threw a party. She told her guests, "I have a riddle for you. I have three daughters. The product of their ages is 72. The sum of their ages is the same as my house number. How old are my daughters?"

The guests went outside to look at the house number. They thought for a few minutes, and then said, "This riddle can't be solved!"

The mathematician said, "Oh yes, I forgot to tell you the last clue. My youngest daughter prefers strawberry ice cream."

With this last clue, the guests could solve the riddle. How old are the mathematician's daughters?

NAME _____ DATE _____ PERIOD _____

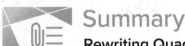

Summary
Rewriting Quadratic Expressions in Factored Form (Part 1)

Previously, you learned how to expand a quadratic expression in factored form and write it in standard form by applying the distributive property.

For example, to expand $(x + 4)(x + 5)$, we apply the distributive property to multiply x by $(x + 5)$ and 4 by $(x + 5)$. Then, we apply the property again to multiply x by x and x by 5, and multiply 4 by x and 4 by 5.

To keep track of all the products, we could make a diagram like this:

	x	4
x		
5		

Next, we could write the products of each pair inside the spaces:

	x	4
x	x^2	$4x$
5	$5x$	$4 \cdot 5$

The diagram helps us see that $(x + 4)(x + 5)$ is equivalent to $x^2 + 5x + 4x + 4 \cdot 5$, or in standard form, $x^2 + 9x + 20$.

- The *linear term*, $9x$, has a *coefficient* of 9, which is the sum of 5 and 4.

- The *constant term*, 20, is the product of 5 and 4.

We can use these observations to reason in the other direction: to start with an expression in standard form and write it in factored form.

For example, suppose we wish to write $x^2 - 11x + 24$ in factored form.

Let's start by creating a diagram and writing in the terms x^2 and 24.

We need to think of two numbers that multiply to make 24 and add up to -11.

	x	
x	x^2	
		24

After some thinking, we see that -8 and -3 meet these conditions.

The product of -8 and -3 is 24. The sum of -8 and -3 is -11.

	x	-8
x	x^2	$-8x$
-3	$-3x$	24

So, $x^2 - 11x + 24$ written in factored form is $(x - 8)(x - 3)$.

Glossary

coefficient
constant term
linear term

---

1. Find two numbers that satisfy the requirements. If you get stuck, try listing all the factors of the first number.

   a. Find two numbers that multiply to 17 and add to 18.

   b. Find two numbers that multiply to 20 and add to 9.

   c. Find two numbers that multiply to 11 and add to -12.

   d. Find two numbers that multiply to 36 and add to -20.

2. Use the diagram to show that:

   $(x + 4)(x + 2)$ is equivalent to $x^2 + 6x + 8$.

   $(x - 10)(x - 3)$ is equivalent to $x^2 - 13x + 30$.

	x	2
x		
4		

	x	-10
x	x^2	-10x
-3	-3x	30

3. Select **all** expressions that are equivalent to $x - 5$.

   (A.) $x + (-5)$

   (B.) $x - (-5)$

   (C.) $-5 + x$

   (D.) $-5 - x$

   (E.) $5 - x$

   (F.) $-5 - (-x)$

   (G.) $5 + x$

4. Here are pairs of equivalent expressions—one in standard form and the other in factored form. Find the missing numbers.

   a. $x^2 + \boxed{}x + \boxed{}$ and $(x - 9)(x - 3)$

   b. $x^2 + 12x + 32$ and $(x + 4)(x + \boxed{})$

   c. $x^2 - 12x + 35$ and $(x - 5)(x + \boxed{})$

   d. $x^2 - 9x + 20$ and $(x - 4)(x + \boxed{})$

NAME _____ DATE _____ PERIOD _____

5. Find all the values for the variable that make each equation true. **(Lesson 7-4)**

  a. $b(b - 4.5) = 0$

  b. $(7x + 14)(7x + 14) = 0$

  c. $(2x + 4)(x - 4) = 0$

  d. $(-2 + u)(3 - u) = 0$

6. Lin charges $5.50 per hour to babysit. The amount of money earned, in dollars, is a function of the number of hours that she babysits.

  Which of the following inputs is impossible for this function? **(Lesson 4-10)**

  (A.) -1

  (B.) 2

  (C.) 5

  (D.) 8

7. Consider the function $p(x) = \dfrac{x - 3}{2x - 6}$. **(Lesson 4-10)**

  a. Evaluate $p(1)$, writing out every step.

  b. Evaluate $p(3)$, writing out every step. You will run into some trouble. Describe it.

  c. What is a possible domain for p?

8. *Technology required.* When solving the equation $(2 - x)(x + 1) = 11$, Priya graphs $y = (2 - x)(x + 1) - 11$ and then looks to find where the graph crosses the *x*-axis. (Lesson 7-5)

Tyler looks at her work and says that graphing is unnecessary and Priya can set up the equations $2 - x = 11$ and $x + 1 = 11$, so the solutions are $x = -9$ or $x = 10$.

a. Do you agree with Tyler? If not, where is the mistake in his reasoning?

b. How many solutions does the equation have? Find out by graphing Priya's equation.

Lesson 7-7

Rewriting Quadratic Expressions in Factored Form (Part 2)

NAME _____ DATE _____ PERIOD _____

Learning Goal Let's write some more expressions in factored form.

 ## Warm Up
7.1 Sums and Products

1. The product of the integers 2 and -6 is -12. List all the other pairs of integers whose product is -12.

2. Of the pairs of factors you found, list all pairs that have a positive sum. Explain why they all have a positive sum.

3. Of the pairs of factors you found, list all pairs that have a negative sum. Explain why they all have a negative sum.

 ## Activity
7.2 Negative Constant Terms

1. These expressions are like the ones we have seen before.

Factored Form	Standard Form
$(x + 5)(x + 6)$	
	$x^2 + 13x + 30$
$(x - 3)(x - 6)$	
	$x^2 - 11x + 18$

Each row has a pair of equivalent expressions.

Complete the table. If you get stuck, consider drawing a diagram.

2. These expressions are in some ways unlike the ones we have seen before.

Factored Form	Standard Form
$(x + 12)(x - 3)$	
	$x^2 - 9x - 36$
	$x^2 - 35x - 36$
	$x^2 + 35x - 36$

Each row has a pair of equivalent expressions.

Complete the table. If you get stuck, consider drawing a diagram.

3. Name some ways that the expressions in the second table are different from those in the first table (aside from the fact that the expressions use different numbers).

Activity

7.3 Factors of 100 and -100

1. Consider the expression $x^2 + bx + 100$.

   Complete the first table with all pairs of factors of 100 that would give positive values of b, and the second table with factors that would give negative values of b.

   For each pair, state the b value they produce. (Use as many rows as needed.)

   positive value of b

Factor 1	Factor 2	b (Positive)

   negative value of b

Factor 1	Factor 2	b (Negative)

NAME _____ DATE _____ PERIOD _____

2. Consider the expression $x^2 + bx - 100$.

Complete the first table with all pairs of factors of -100 that would result in positive values of b, the second table with factors that would result in negative values of b, and the third table with factors that would result in a zero value of b.

For each pair of factors, state the b value they produce. (Use as many rows as there are pairs of factors. You may not need all the rows.)

positive value of b

Factor 1	Factor 2	b (Positive)

negative value of b

Factor 1	Factor 2	b (Negative)

zero value of b

Factor 1	Factor 2	b (Zero)

3. Write each expression in factored form:

a. $x^2 - 25x + 100$

b. $x^2 + 15x - 100$

c. $x^2 - 15x - 100$

d. $x^2 + 99x - 100$

How many different integers b can you find so that the expression $x^2 + 10x + b$ can be written in factored form?

NAME _____ DATE _____ PERIOD _____

Summary
Rewriting Quadratic Expressions in Factored Form (Part 2)

When we rewrite expressions in factored form, it is helpful to remember that:

- Multiplying two positive numbers or two negative numbers results in a positive product.

- Multiplying a positive number and a negative number results in a negative product.

This means that if we want to find two factors whose product is 10, the factors must be both positive or both negative. If we want to find two factors whose product is -10, one of the factors must be positive and the other negative.

Suppose we wanted to rewrite $x^2 - 8x + 7$ in factored form. Recall that subtracting a number can be thought of as adding the opposite of that number, so that expression can also be written as $x^2 + -8x + 7$. We are looking for two numbers that:

- Have a product of 7. The candidates are 7 and 1, and -7 and -1.

- Have a sum of -8. Only -7 and -1 from the list of candidates meet this condition.

The factored form of $x^2 - 8x + 7$ is therefore $(x + -7)(x + -1)$ or, written another way, $(x - 7)(x - 1)$.

To write $x^2 + 6x - 7$ in factored form, we would need two numbers that:

- Multiply to make -7. The candidates are 7 and -1, and -7 and 1.

- Add up to 6. Only 7 and -1 from the list of candidates add up to 6.

The factored form of $x^2 + 6x - 7$ is $(x + 7)(x - 1)$.

Practice
Rewriting Quadratic Expressions in Factored Form (Part 2)

1. Find two numbers that. . .

   a. multiply to -40 and add to -6.

   b. multiply to -40 and add to 6.

   c. multiply to -36 and add to 9.

   d. multiply to -36 and add to -5.

   If you get stuck, try listing all the factors of the first number.

2. Create a diagram to show that $(x - 5)(x + 8)$ is equivalent to $x^2 + 3x - 40$.

3. Write a + or a − sign in each box so the expressions on each side of the equal sign are equivalent.

   a. $(x \boxed{} 18)(x \boxed{} 3) = x^2 - 15x - 54$

   b. $(x \boxed{} 18)(x \boxed{} 3) = x^2 + 21x + 54$

   c. $(x \boxed{} 18)(x \boxed{} 3) = x^2 + 15x - 54$

   d. $(x \boxed{} 18)(x \boxed{} 3) = x^2 - 21x + 54$

4. Match each quadratic expression in standard form with its equivalent expression in factored form.

   A. $x^2 - 2x - 35$ 1. $(x + 5)(x + 7)$

   B. $x^2 + 12x + 35$ 2. $(x - 5)(x - 7)$

   C. $x^2 + 2x - 35$ 3. $(x + 5)(x - 7)$

   D. $x^2 - 12x + 35$ 4. $(x - 5)(x + 7)$

NAME _____ DATE _____ PERIOD _____

5. Rewrite each expression in factored form. If you get stuck, try drawing a diagram.

   a. $x^2 - 3x - 28$

   b. $x^2 + 3x - 28$

   c. $x^2 + 12x - 28$

   d. $x^2 - 28x - 60$

6. Which equation has exactly one solution? **(Lesson 7-5)**

   (A.) $x^2 = -4$

   (B.) $(x + 5)^2 = 0$

   (C.) $(x + 5)(x - 5) = 0$

   (D.) $(x + 5)^2 = 36$

7. The graph represents the height of a passenger car on a ferris wheel, in feet, as a function of time, in seconds since the ride starts. **(Lesson 4-11)**

   Use the graph to help you:

   a. Find $H(0)$.

   b. Does $H(t) = 0$ have a solution? Explain how you know.

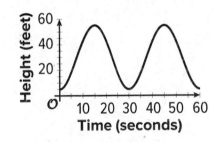

   c. Describe the domain of the function.

   d. Describe the range of the function.

8. Elena solves the equation $x^2 = 7x$ by dividing both sides by x to get $x = 7$. She says the solution is 7. (Lesson 7-5)

Lin solves the equation $x^2 = 7x$ by rewriting the equation to get $x^2 - 7x = 0$. When she graphs the equation $y = x^2 - 7x$, the x-intercepts are (0, 0) and (7, 0). She says the solutions are 0 and 7.

Do you agree with either of them? Explain or show how you know.

9. A bacteria population, p, can be represented by the equation $p = 100,000 \cdot \left(\frac{1}{4}\right)^d$, where d is the number of days since it was measured. (Lesson 5-7)

   a. What was the population 3 days before it was measured? Explain how you know.

   b. What is the last day when the population was more than 1,000,000? Explain how you know.

Lesson 7-8

Rewriting Quadratic Expressions in Factored Form (Part 3)

NAME _____ DATE _____ PERIOD _____

Learning Goal Let's look closely at some special kinds of factors.

Warm Up
8.1 Math Talk: Products of Large-ish Numbers

Find each product mentally.

$9 \cdot 11$

$19 \cdot 21$

$99 \cdot 101$

$109 \cdot 101$

Activity
8.2 Can Products Be Written as Differences?

1. Clare claims that $(10 + 3)(10 - 3)$ is equivalent to $10^2 - 3^2$ and $(20 + 1)(20 - 1)$ is equivalent to $20^2 - 1^2$. Do you agree? Show your reasoning.

2. Respond to each question.

   a. Use your observations from the first question and evaluate $(100 + 5)(100 - 5)$. Show your reasoning.

   b. Check your answer by computing $105 \cdot 95$.

3. Is $(x + 4)(x - 4)$ equivalent to $x^2 - 4^2$? Support your answer:

With a diagram:

	x	4
x		
-4		

Without a diagram:

4. Is $(x + 4)^2$ equivalent to $x^2 + 4^2$? Support your answer, either with or without a diagram.

Are you ready for more?

1. Explain how your work in the previous questions can help you mentally evaluate $22 \cdot 18$ and $45 \cdot 35$.

2. Here is a shortcut that can be used to mentally square any two-digit number. Let's take 83^2, for example.

   - 83 is $80 + 3$.

   - Compute 80^2 and 3^2, which give 6,400 and 9. Add these values to get 6,409.

   - Compute $80 \cdot 3$, which is 240. Double it to get 480.

   - Add 6,409 and 480 to get 6,889.

   Try using this method to find the squares of some other two-digit numbers. (With some practice, it is possible to get really fast at this!) Then, explain why this method works.

NAME _____ DATE _____ PERIOD _____

Activity

8.3 What If There is No Linear Term?

Each row has a pair of equivalent expressions.

Complete the table.

If you get stuck, consider drawing a diagram. (Heads up: one of them is impossible.)

Factored Form	Standard Form
$(x - 10)(x + 10)$	
$(2x + 1)(2x - 1)$	
$(4 - x)(4 + x)$	
	$x^2 - 81$
	$49 - y^2$
	$9z^2 - 16$
	$25t^2 - 81$
$\left(c + \frac{2}{5}\right)\left(c - \frac{2}{5}\right)$	
	$\frac{49}{16} - d^2$
$(x + 5)(x + 5)$	
	$x^2 - 6$
	$x^2 + 100$

Summary
Rewriting Quadratic Expressions in Factored Form (Part 3)

Sometimes expressions in standard form don't have a linear term. Can they still be written in factored form?

Let's take $x^2 - 9$ as an example. To help us write it in factored form, we can think of it as having a linear term with a coefficient of 0: $x^2 + 0x - 9$. (The expression $x^2 - 0x - 9$ is equivalent to $x^2 - 9$ because 0 times any number is 0, so $0x$ is 0.)

We know that we need to find two numbers that multiply to make -9 and add up to 0. The numbers 3 and -3 meet both requirements, so the factored form is $(x + 3)(x - 3)$.

To check that this expression is indeed equivalent to $x^2 - 9$, we can expand the factored expression by applying the distributive property: $(x + 3)(x - 3)$ $= x^2 - 3x + 3x + (-9)$. Adding $-3x$ and $3x$ gives 0, so the expanded expression is $x^2 - 9$.

In general, a quadratic expression that is a difference of two squares and has the form:

$a^2 - b^2$ can be rewritten as: $(a + b)(a - b)$

Here is a more complicated example: $49 - 16y^2$. This expression can be written $7^2 - (4y)^2$, so an equivalent expression in factored form is $(7 + 4y)$ $(7 - 4y)$.

What about $x^2 + 9$? Can it be written in factored form?

Let's think about this expression as $x^2 + 0x + 9$. Can we find two numbers that multiply to make 9 but add up to 0? Here are factors of 9 and their sums:

- 9 and 1, sum: 10

- -9 and -1, sum: -10

- 3 and 3, sum: 6

- -3 and -3, sum: -6

For two numbers to add up to 0, they need to be opposites (a negative and a positive), but a pair of opposites cannot multiply to make positive 9, because multiplying a negative number and a positive number always gives a negative product.

Because there are no numbers that multiply to make 9 and also add up to 0, it is not possible to write $x^2 + 9$ in factored form using the kinds of numbers that we know about.

NAME _____ DATE _____ PERIOD _____

Practice
Rewriting Quadratic Expressions in Factored Form (Part 3)

1. Match each quadratic expression given in factored form with an equivalent expression in standard form. One expression in standard form has no match.

   A. $(y + x)(y - x)$

   B. $(11 + x)(11 - x)$

   C. $(x - 11)(x + 11)$

   D. $(x - y)(x - y)$

   1. $121 - x^2$

   2. $x^2 + 2xy - y^2$

   3. $y^2 - x^2$

   4. $x^2 - 2xy + y^2$

   5. $x^2 - 121$

2. Both $(x - 3)(x + 3)$ and $(3 - x)(3 + x)$ contain a sum and a difference and have only 3 and x in each factor.

   If each expression is rewritten in standard form, will the two expressions be the same? Explain or show your reasoning.

3. Respond to each question.

   a. Show that the expressions $(5 + 1)(5 - 1)$ and $5^2 - 1^2$ are equivalent.

   b. The expressions $(30 - 2)(30 + 2)$ and $30^2 - 2^2$ are equivalent and can help us find the product of two numbers. Which two numbers are they?

   c. Write $94 \cdot 106$ as a product of a sum and a difference, and then as a difference of two squares. What is the value of $94 \cdot 106$?

4. Write each expression in factored form. If not possible, write "not possible."

   a. $x^2 - 144$

   b. $x^2 + 16$

   c. $25 - x^2$

   d. $b^2 - a^2$

   e. $100 + y^2$

5. What are the solutions to the equation $(x - a)(x + b) = 0$? (Lesson 7-4)

   (A.) a and b

   (B.) $-a$ and $-b$

   (C.) a and $-b$

   (D.) $-a$ and b

6. Create a diagram to show that $(x - 3)(x - 7)$ is equivalent to $x^2 - 10x + 21$.
   (Lesson 7-6)

7. Select **all** the expressions that are equivalent to $8 - x$. (Lesson 7-6)

   (A.) $x - 8$

   (B.) $8 + (-x)$

   (C.) $-x - (-8)$

   (D.) $-8 + x$

   (E.) $x - (-8)$

   (F.) $x + (-8)$

   (G.) $-x + 8$

NAME _____ DATE _____ PERIOD _____

8. Mai fills a tall cup with hot cocoa, 12 centimeters in height. She waits
   5 minutes for it to cool. Then, she starts drinking in sips, at an average rate
   of 2 centimeters of height every 2 minutes, until the cup is empty.

   The function C gives the height of hot cocoa in Mai's cup, in centimeters, as
   a function of time, in minutes. **(Lesson 4-11)**

   a. Sketch a possible graph of C. Be sure to include a label and a scale for
      each axis.

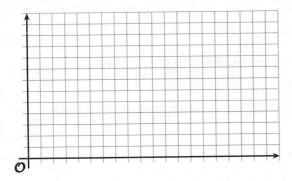

   b. What quantities do the domain and range represent in this situation?

   c. Describe the domain and range of C.

9. Two bacteria populations are measured at the same time. One bacteria population, p, is modeled by the equation $p = 250{,}000 \cdot \left(\frac{1}{2}\right)^d$, where d is the number of days since it was first measured.

A second bacteria population, q, is modeled by the equation $q = 500{,}000 \cdot \left(\frac{1}{3}\right)^d$, where d is the number of days since it was first measured.

Which statement is true about the two populations? (Lesson 5-7)

(A.) The second population will always be larger than the first.

(B.) Both populations are increasing.

(C.) The second bacteria population decreases more rapidly than the first.

(D.) When initially measured, the first population is larger than the second.

Lesson 7-9

Solving Quadratic Equations by Using Factored Form

NAME _____ DATE _____ PERIOD _____

Learning Goal Let's solve some quadratic equations that before now we could only solve by graphing.

 ## Warm Up
9.1 Why Would You Do That?

Let's try to find at least one solution to $x^2 - 2x - 35 = 0$.

1. Choose a whole number between 0 and 10.

2. Evaluate the expression $x^2 - 2x - 35$, using your number for x.

3. If your number doesn't give a value of 0, look for someone in your class who may have chosen a number that does make the expression equal 0. Which number is it?

4. There is another number that would make the expression $x^2 - 2x - 35$ equal 0. Can you find it?

Activity

9.2 Let's Solve Some Equations!

1. To solve the equation $n^2 - 2n = 99$, Tyler wrote out the following steps. Analyze Tyler's work. Write down what Tyler did in each step.

$$n^2 - 2n = 99 \qquad \text{Original equation}$$
$$n^2 - 2n - 99 = 0 \qquad \text{Step 1}$$
$$(n - 11)(n + 9) = 0 \qquad \text{Step 2}$$
$$n - 11 = 0 \ \text{ or } \ n + 9 = 0 \qquad \text{Step 3}$$
$$n = 11 \ \text{ or } \ n = -9 \qquad \text{Step 4}$$

2. Solve each equation by rewriting it in factored form and using the zero product property. Show your reasoning.

   a. $x^2 + 8x + 15 = 0$

   b. $x^2 - 8x + 12 = 5$

   c. $x^2 - 10x - 11 = 0$

   d. $49 - x^2 = 0$

   e. $(x + 4)(x + 5) - 30 = 0$

Are you ready for more?

Solve this equation and explain or show your reasoning.

$$(x^2 - x - 20)(x^2 + 2x - 3) = (x^2 + 2x - 8)(x^2 - 8x + 15)$$

NAME _____ DATE _____ PERIOD _____

 Activity

9.3 **Revisiting Quadratic Equations with Only One Solution**

1. The other day, we saw that a quadratic equation can have 0, 1, or 2 solutions. Sketch graphs that represent three quadratic functions: one that has no zeros, one with 1 zero, and one with 2 zeros.

   No zeros   **One zero**   **Two zeros**

2. Use graphing technology to graph the function defined by $f(x) = x^2 - 2x + 1$. What do you notice about the x-intercepts of the graph? What do the x-intercepts reveal about the function?

3. Solve $x^2 - 2x + 1 = 0$ by using the factored form and zero product property. Show your reasoning. What solutions do you get?

4. Write an equation to represent another quadratic function that you think will only have one zero. Graph it to check your prediction.

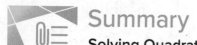

Summary

Solving Quadratic Equations by Using Factored Form

Recently, we learned strategies for transforming expressions from standard form to factored form. In earlier lessons, we have also seen that when a quadratic expression is in factored form, it is pretty easy to find values of the variable that make the expression equal zero. Suppose we are solving the equation $x(x + 4) = 0$, which says that the product of x and $x + 4$ is 0. By the zero product property, we know this means that either $x = 0$ or $x + 4 = 0$, which then tells us that 0 and -4 are solutions.

Together, these two skills—writing quadratic expressions in factored form and using the zero product property when a factored expression equals 0—allow us to solve quadratic equations given in other forms. Here is an example:

$n^2 - 4n = 140$	Original equation
$n^2 - 4n - 140 = 0$	Subtract 140 from each side so the right side is 0
$(n - 14)(n + 10) = 0$	Rewrite in factored form
$n - 14 = 0$ or $n + 10 = 0$	Apply the zero product property
$n = 14$ or $n = -10$	Solve each equation

When a quadratic equation is written as expression in factored form = 0, we can also see the number of solutions the equation has.

In the example earlier, it was not obvious how many solutions there would be when the equation was $n^2 - 4n - 140 = 0$. When the equation was rewritten as $(n - 14)(n + 10) = 0$, we could see that there were two numbers that could make the expression equal 0: 14 and -10.

How many solutions does the equation $x^2 - 20x + 100 = 0$ have?

Let's rewrite it in factored form: $(x - 10)(x - 10) = 0$. The two factors are identical, which means that there is only one value of x that makes the expression $(x - 10)(x - 10)$ equal 0. The equation has only one solution: 10.

NAME _____ DATE _____ PERIOD _____

Practice
Solving Quadratic Equations by Using Factored Form

1. Find **all** the solutions to each equation.

   a. $x(x - 1) = 0$

   b. $(5 - x)(5 + x) = 0$

   c. $(2x + 1)(x + 8) = 0$

   d. $(3x - 3)(3x - 3) = 0$

   e. $(7 - x)(x + 4) = 0$

2. Rewrite each equation in factored form and solve using the zero product property.

   a. $d^2 - 7d + 6 = 0$

   b. $x^2 + 18x + 81 = 0$

   c. $u^2 + 7u - 60 = 0$

   d. $x^2 + 0.2x + 0.01 = 0$

3. Here is how Elena solves the quadratic equation $x^2 - 3x - 18 = 0$.

$x^2 - 3x - 18 = 0$

$(x - 3)(x + 6) = 0$

$x - 3 = 0$ or $x + 6 = 0$

$x = 3$ or $x = -6$

Is her work correct? If you think there is an error, explain the error and correct it.

Otherwise, check her solutions by substituting them into the original equation and showing that the equation remains true.

4. Jada is working on solving a quadratic equation, as shown here.

$p^2 - 5p = 0$

$p(p - 5) = 0$

$p - 5 = 0$

$p = 5$

She thinks that her solution is correct because substituting 5 for p in the original expression $p^2 - 5p$ gives $5^2 - 5(5)$, which is $25 - 25$ or 0.

Explain the mistake that Jada made and show the correct solutions.

5. Choose a statement to correctly describe the zero product property.

   If a and b are numbers, and $a \cdot b = 0$, then: (Lesson 7-4)

   (A.) Both a and b must equal 0.

   (B.) Neither a nor b can equal 0.

   (C.) Either $a = 0$ or $b = 0$.

   (D.) $a + b$ must equal 0.

6. Which expression is equivalent to $x^2 - 7x + 12$? (Lesson 7-6)

   (A.) $(x + 3)(x + 4)$

   (B.) $(x - 3)(x - 4)$

   (C.) $(x + 2)(x + 6)$

   (D.) $(x - 2)(x - 6)$

7. These quadratic expressions are given in standard form. Rewrite each expression in factored form. If you get stuck, try drawing a diagram. (Lesson 7-6)

   a. $x^2 + 7x + 6$

   b. $x^2 - 7x + 6$

   c. $x^2 - 5x + 6$

   d. $x^2 + 5x + 6$

8. Select **all** the functions whose output values will eventually overtake the output values of function f defined by $f(x) = 25x^2$. (Lesson 6-4)

   (A.) $g(x) = 5(2)^x$

   (B.) $h(x) = 5^x$

   (C.) $j(x) = x^2 + 5$

   (D.) $k(x) = \left(\frac{5}{2}\right)^x$

   (E.) $m(x) = 5 + 2^x$

   (F.) $n(x) = 2x^2 + 5$

9. A piecewise function, p, is defined by this rule: $p(x) = \begin{cases} x - 1, & x \le \text{-}2 \\ 2x - 1, & x > \text{-}2 \end{cases}$

   Find the value of p at each given input. (Lesson 4-12)

   a. $p(\text{-}20)$

   b. $p(\text{-}2)$

   c. $p(4)$

   d. $p(5.7)$

Lesson 7-10

Rewriting Quadratic Expressions in Factored Form (Part 4)

NAME _____ DATE _____ PERIOD _____

Learning Goal Let's transform more-complicated quadratic expressions into the factored form.

 ## Warm Up

10.1 Which One Doesn't Belong: Quadratic Expressions

Which one doesn't belong?

A. $(x + 4)(x - 3)$

B. $3x^2 - 8x + 5$

C. $x^2 - 25$

D. $x^2 + 2x + 3$

Each row in each table has a pair of equivalent expressions. Complete the tables. If you get stuck, try drawing a diagram.

1.

Factored Form	Standard Form
$(3x + 1)(x + 4)$	
$(3x + 2)(x + 2)$	
$(3x + 4)(x + 1)$	

2.

Factored Form	Standard Form
	$5x^2 + 21x + 4$
	$3x^2 + 15x + 12$
	$6x^2 + 19x + 10$

Are you ready for more?

Here are three quadratic equations, each with two solutions. Find both solutions to each equation, using the zero product property somewhere along the way. Show each step in your reasoning.

$$x^2 = 6x \qquad x(x + 4) = x + 4 \qquad 2x(x - 1) + 3x - 3 = 0$$

NAME _____ DATE _____ PERIOD _____

Activity

10.3 Timing A Blob of Water

An engineer is designing a fountain that shoots out drops of water. The nozzle from which the water is launched is 3 meters above the ground. It shoots out a drop of water at a vertical velocity of 9 meters per second.

Function h models the height in meters, h, of a drop of water t seconds after it is shot out from the nozzle. The function is defined by the equation $h(t) = -5t^2 + 9t + 3$.

How many seconds until the drop of water hits the ground?

1. Write an equation that we could solve to answer the question.

2. Try to solve the equation by writing the expression in factored form and using the zero product property.

3. Try to solve the equation by graphing the function using graphing technology. Explain how you found the solution.

Here is a clever way to think about quadratic expressions that would make it easier to rewrite them in factored form.

$$9x^2 + 21x + 10$$
$$(3x)^2 + 7(3x) + 10$$
$$N^2 + 7N + 10$$
$$(N + 2)(N + 5)$$
$$(3x + 2)(3x + 5)$$

1. Use the distributive property to expand $(3x + 2)(3x + 5)$. Show your reasoning and write the resulting expression in standard form. Is it equivalent to $9x^2 + 21x + 10$?

2. Study the method and make sense of what was done in each step. Make a note of your thinking and be prepared to explain it.

3. Try the method to write each of these expressions in factored form.

$$4x^2 + 28x + 45 \qquad\qquad 25x^2 - 35x + 6$$

NAME _____ DATE _____ PERIOD _____

4. You have probably noticed that the coefficient of the squared term in all of the previous examples is a perfect square. What if that coefficient is not a perfect square?

Here is an example of an expression whose squared term has a coefficient that is not a squared term.

$$5x^2 + 17x + 6$$

$$\frac{1}{5} \cdot 5 \cdot (5x^2 + 17x + 6)$$

$$\frac{1}{5}(25x^2 + 85x + 30)$$

$$\frac{1}{5}((5x)^2 + 17(5x) + 30)$$

$$\frac{1}{5}(N^2 + 17N + 30)$$

$$\frac{1}{5}(N + 15)(N + 2)$$

$$\frac{1}{5}(5x + 15)(5x + 2)$$

$$(x + 3)(5x + 2)$$

Use the distributive property to expand $(x + 3)(5x + 2)$. Show your reasoning and write the resulting expression in standard form. Is it equivalent to $5x^2 + 17x + 6$?

5. Study the method and make sense of what was done in each step and why. Make a note of your thinking and be prepared to explain it.

6. Try the method to write each of these expressions in factored form.

$$3x^2 + 16x + 5 \qquad\qquad 10x^2 - 41x + 4$$

Summary
Rewriting Quadratic Expressions in Factored Form (Part 4)

Only some quadratic equations in the form of $ax^2 + bx + c = 0$ can be solved by rewriting the quadratic expression into factored form and using the zero product property. In some cases, finding the right factors of the quadratic expression is quite difficult.

For example, what is the factored form of $6x^2 + 11x - 35$?

We know that it could be $(3x + \boxed{})(2x + \boxed{})$, or $(6x + \boxed{})(x + \boxed{})$, but will the second number in each factor be -5 and 7, 5 and -7, 35 and -1, or -35 and 1? And in which order?

We have to do some guessing and checking before finding the equivalent expression that would allow us to solve the equation $6x^2 + 11x - 35 = 0$.

Once we find the right factors, we can proceed to solving using the zero product property, as shown here:

$$6x^2 + 11x - 35 = 0$$
$$(3x - 5)(2x + 7) = 0$$
$$3x - 5 = 0 \quad \text{or} \quad 2x + 7 = 0$$
$$x = \frac{5}{3} \quad \text{or} \quad x = -\frac{7}{2}$$

What is even trickier is that most quadratic expressions can't be written in factored form!

Let's take $x^2 - 4x - 3$ for example. Can you find two numbers that multiply to make -3 and add up to -4? Nope! At least not easy-to-find rational numbers.

NAME _____ DATE _____ PERIOD _____

We can graph the function defined by $x^2 - 4x - 3$ using technology, which reveals two x-intercepts, at around (-0.646, 0) and (4.646, 0). These give the approximate zeros of the function, -0.646 and 4.646, so they are also approximate solutions to $x^2 - 4x - 3 = 0$.

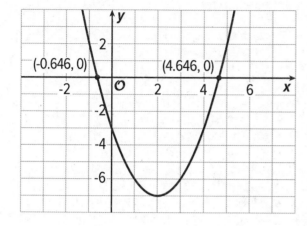

The fact that the zeros of this function don't seem to be simple rational numbers is a clue that it may not be possible to easily rewrite the expression in factored form.

It turns out that rewriting quadratic expressions in factored form and using the zero product property is a very limited tool for solving quadratic equations.

In the next several lessons, we will learn some ways to solve quadratic equations that work for any equation.

Practice

Rewriting Quadratic Expressions in Factored Form (Part 4)

1. To write $11x^2 + 17x - 10$ in factored form, Diego first listed pairs of factors of -10.

   (____ + 5)(____ + -2)

   (____ + 2)(____ + -5)

   (____ + 10)(____ + -1)

   (____ + 1)(____ + -10)

   a. Use what Diego started to complete the rewriting.

   b. How did you know you've found the right pair of expressions? What did you look for when trying out different possibilities?

2. To rewrite $4x^2 - 12x - 7$ in factored form, Jada listed some pairs of factors of $4x^2$:

   (2x + ____)(2x + ____)

   (4x + ____)(1x + ____)

   Use what Jada started to rewrite $4x^2 - 12x - 7$ in factored form.

3. Rewrite each quadratic expression in factored form. Then, use the zero product property to solve the equation.

   a. $7x^2 - 22x + 3 = 0$

   b. $4x^2 + x - 5 = 0$

   c. $9x^2 - 25 = 0$

4. Han is solving the equation $5x^2 + 13x - 6 = 0$. Here is his work:

   $5x^2 + 13x - 6 = 0$

   $(5x - 2)(x + 3) = 0$

   $x = 2$ or $x = -3$

   Describe Han's mistake. Then, find the correct solutions to the equation.

NAME _____ DATE _____ PERIOD _____

5. A picture is 10 inches wide by 15 inches long. The area of the picture, including a frame that is x inch thick, can be modeled by the function $A(x) = (2x + 10)(2x + 15)$. (Lesson 7-1)

   a. Use function notation to write a statement that means: the area of the picture, including a frame that is 2 inches thick, is 266 square inches.

   b. What is the total area if the picture has a frame that is 4 inches thick?

6. To solve the equation $0 = 4x^2 - 28x + 39$, Elena uses technology to graph the function $f(x) = 4x^2 - 28x + 39$. She finds that the graph crosses the x-axis at (1.919, 0) and (5.081, 0). (Lesson 7-2)

   a. What is the name for the points where the graph of a function crosses the x-axis?

   b. Use a calculator to compute $f(1.919)$ and $f(5.081)$.

   c. Explain why 1.919 and 5.081 are approximate solutions to the equation $0 = 4x^2 - 28x + 39$ and are not exact solutions.

7. Which equation shows a next step in solving $9(x - 1)^2 = 36$ that will lead to the correct solutions? (Lesson 7-3)

   Ⓐ $9(x - 1) = 6$ or $9(x - 1) = -6$

   Ⓑ $3(x - 1) = 6$

   Ⓒ $(x - 1)^2 = 4$

   Ⓓ $(9x - 9)^2 = 36$

8. Here is a description of the temperature at a certain location yesterday.

   "It started out cool in the morning, but then the temperature increased until noon. It stayed the same for a while, until it suddenly dropped quickly! It got colder than it was in the morning, and after that, it was cold for the rest of the day."

   Sketch a graph of the temperature as a function of time. (Lesson 4-8)

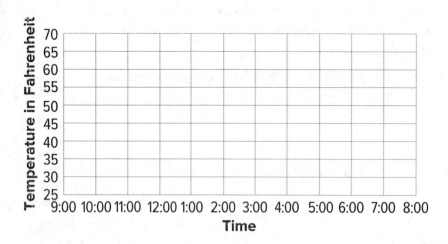

9. *Technology required.* The number of people, p, who watch a weekly TV show is modeled by the equation $p = 100{,}000 \cdot (1.1)^w$, where w is the number of weeks since the show first aired. (Lesson 5-9)

   a. How many people watched the show the first time it aired? Explain how you know.

   b. Use technology to graph the equation.

   c. On which week does the show first get an audience of more than 500,000 people?

Lesson 7-11

What are Perfect Squares?

NAME _____ DATE _____ PERIOD _____

Learning Goal Let's see how perfect squares make some equations easier to solve.

 ## Warm Up
11.1 The Thing We Are Squaring

In each equation, what expression could be substituted for a so the equation is true for all values of x?

1. $x^2 = a^2$

2. $(3x)^2 = a^2$

3. $a^2 = 7x \cdot 7x$

4. $25x^2 = a^2$

5. $a^2 = \frac{1}{4}x^2$

6. $a^2 = (x + 1)^2$

7. $(2x - 9)(2x - 9) = a^2$

 ## Activity
11.2 Perfect Squares in Different Forms

1. Each expression is written as the product of factors. Write an equivalent expression in standard form.

   a. $(3x)^2$

   b. $7x \cdot 7x$

   c. $(x + 4)(x + 4)$

   d. $(x + 1)^2$

   e. $(x - 7)^2$

   f. $(x + n)^2$

2. Why do you think the following expressions can be described as **perfect squares**?

$$x^2 + 6x + 9 \qquad x^2 - 16x + 64 \qquad x^2 + \frac{1}{3}x + \frac{1}{36}$$

Are you ready for more?

Write each expression in factored form.

1. $x^4 - 30x^2 + 225$

2. $x + 14\sqrt{x} + 49$

3. $5^{2x} + 6 \cdot 5^x + 9$

 ## Activity
11.3 Two Methods

Han and Jada solved the same equation with different methods. Here they are:

Han's method:

$$(x - 6)^2 = 25$$
$$(x - 6)(x - 6) = 25$$
$$x^2 - 12x + 36 = 25$$
$$x^2 - 12x + 11 = 0$$
$$(x - 11)(x - 1) = 0$$

$$x = 11 \quad \text{or} \quad x = 1$$

Jada's method:

$$(x - 6)^2 = 25$$

$$x - 6 = 5 \quad \text{or} \quad x - 6 = -5$$
$$x = 11 \quad \text{or} \quad x = 1$$

Work with a partner to solve these equations. For each equation, one partner solves with Han's method, and the other partner solves with Jada's method. Make sure both partners get the same solutions to the same equation. If not, work together to find your mistakes.

$$(y - 5)^2 = 49 \qquad\qquad\qquad (x + 4)^2 = 9$$

$$\left(z + \frac{1}{3}\right)^2 = \frac{4}{9} \qquad\qquad\qquad (v - 0.1)^2 = 0.36$$

NAME _____ DATE _____ PERIOD _____

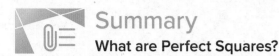

Summary
What are Perfect Squares?

These are some examples of **perfect squares**:

- 49, because 49 is $7 \cdot 7$ or 7^2.
- $81a^2$, because it is equivalent to $(9a) \cdot (9a)$ or $(9a)^2$.
- $(x + 5)^2$, because it is equivalent to $(x + 5)(x + 5)$.
- $x^2 - 12x + 36$, because it is equivalent to $(x - 6)^2$ or $(x - 6)(x - 6)$.

A *perfect square* is an expression that is something times itself. Usually we are interested in situations in which the something is a *rational* number or an expression with rational coefficients.

When expressions that are perfect squares are written in factored form and standard form, there is a predictable pattern.

- $(x + 5)(x + 5)$ is equivalent to $x^2 + 10x + 25$.
- $(x - 6)^2$ is equivalent to $x^2 - 12x + 36$.
- $(x - 9)^2$ is equivalent to $x^2 - 18x + 81$.

In general, $(x + n)^2$ is equivalent to $x^2 + (2n)x + n^2$.

Quadratic equations that are in the form a perfect square = a perfect square can be solved in a straightforward manner. Here is an example:

$$
\begin{aligned}
x^2 - 18x + 81 &= 25 \\
(x - 9)(x - 9) &= 25 \\
(x - 9)^2 &= 25
\end{aligned}
$$

The equation now says: squaring $(x - 9)$ gives 25 as a result. This means $(x - 9)$ must be 5 or -5.

$$
\begin{array}{ccc}
x - 9 = 5 & \text{or} & x - 9 = -5 \\
x = 14 & \text{or} & x = 4
\end{array}
$$

Glossary

perfect square
rational number

Practice

What are Perfect Squares?

1. Select **all** the expressions that are perfect squares.

   (A.) $(x + 5)(x + 5)$ (D.) $(a + 3)(3 + a)$ (F.) $(4 - 3x)(3 - 4x)$

   (B.) $(-9 + c)(c - 9)$ (E.) $(2x - 1)(2x + 1)$ (G.) $(a + b)(b + a)$

   (C.) $(y - 10)(10 - y)$

2. Each diagram represents the square of an expression or a perfect square.

   $(n + 7)^2$

	n	7
n	n^2	$7n$
7	$7n$	7^2

   $(5 - m)^2$

	5	$-m$
5	5^2	$5(-m)$
$-m$	$5(-m)$	$(-m)^2$

   $\left(h + \dfrac{1}{3}\right)^2$

	h	$\dfrac{1}{3}$
h		
$\dfrac{1}{3}$		

   a. Complete the cells in the last table.

   b. How are the contents of the three diagrams alike? This diagram represents $(\text{term\_1} + \text{term\_2})^2$. Describe your observations about cells 1, 2, 3, and 4.

	Term_1	Term_2
Term_1	cell 1	cell 2
Term_2	cell 3	cell 4

   c. Rewrite the perfect-square expressions $(n + 7)^2$, $(5 - m)^2$, and $\left(h + \dfrac{1}{3}\right)^2$ in standard form: $ax^2 + bx + c$.

   d. How are the ax^2, bx, and c of a perfect square in standard form related to the two terms in $(\text{term\_1} + \text{term\_2})^2$?

NAME _____ DATE _____ PERIOD _____

3. Solve each equation.

  a. $(x - 1)^2 = 4$

  b. $(x + 5)^2 = 81$

  c. $(x - 2)^2 = 0$

  d. $(x + 11)^2 = 121$

  e. $(x - 7)^2 = \dfrac{64}{49}$

4. Explain or show why the product of a sum and a difference, such as $(2x + 1)(2x - 1)$, has no linear term when written in standard form. **(Lesson 7-8)**

5. To solve the equation $(x + 3)^2 = 4$, Han first expanded the squared expression. Here is his incomplete work:

$$(x + 3)^2 = 4$$
$$(x + 3)(x + 3) = 4$$
$$x^2 + 3x + 3x + 9 = 4$$
$$x^2 + 6x + 9 = 4$$

   a. Complete Han's work and solve the equation.

   b. Jada saw the equation $(x + 3)^2 = 4$ and thought, "There are two numbers, 2 and -2, that equal 4 when squared. This means $x + 3$ is either 2 or it is -2. I can find the values of x from there."

   Use Jada's reasoning to solve the equation.

   c. Can Jada use her reasoning to solve $(x + 3)(x - 3) = 5$? Explain your reasoning.

6. A jar full of marbles is displayed. The following table shows the guesses for 10 people. The actual number of marbles in the jar is 145. Calculate the absolute guessing error for all 10 guesses. (Lesson 4-13)

Guess	190	150	125	133	167	160	148	200	170	115
Absolute Guessing Error										

Lesson 7-12

Completing the Square (Part 1)

DATE _____ PERIOD _____

NAME _____

Learning Goal Let's learn a new method for solving quadratic equations.

Warm Up
12.1 Perfect or Imperfect?

Select **all** expressions that are perfect squares. Explain how you know.

1. $(x + 5)(5 + x)$

2. $(x + 5)(x - 5)$

3. $(x - 3)^2$

4. $x - 3^2$

5. $x^2 + 8x + 16$

6. $x^2 + 10x + 20$

Activity
12.2 Building Perfect Squares

Complete the table so that each row has equivalent expressions that are perfect squares.

Standard Form	Factored Form
1. $x^2 + 6x + 9$	
2. $x^2 - 10x + 25$	
3.	$(x - 7)^2$
4. $x^2 - 20x + $	$(x - \quad)^2$
5. $x^2 + 16x + $	$(x + \quad)^2$
6. $x^2 + 7x + $	$(x + \quad)^2$
7. $x^2 + bx + $	$(x + \quad)^2$

Activity

12.3 Dipping Our Toes in Completing the Square

One technique for solving quadratic equations is called **completing the square**. Here are two examples of how Diego and Mai completed the square to solve the same equation.

Diego:

$$x^2 + 10x + 9 = 0$$
$$x^2 + 10x = -9$$
$$x^2 + 10x + 25 = -9 + 25$$
$$x^2 + 10x + 25 = 16$$
$$(x + 5)^2 = 16$$
$$x + 5 = 4 \quad \text{or} \quad x + 5 = -4$$
$$x = -1 \quad \text{or} \quad x = -9$$

Mai:

$$x^2 + 10x + 9 = 0$$
$$x^2 + 10x + 9 + 16 = 16$$
$$x^2 + 10x + 25 = 16$$
$$(x + 5)^2 = 16$$
$$x + 5 = 4 \quad \text{or} \quad x + 5 = -4$$
$$x = -1 \quad \text{or} \quad x = -9$$

Study the worked examples. Then, try solving these equations by completing the square:

1. $x^2 + 6x + 8 = 0$

2. $x^2 + 12x = 13$

3. $0 = x^2 - 10x + 21$

4. $x^2 - 2x + 3 = 83$

5. $x^2 + 40 = 14x$

Are you ready for more?

Here is a diagram made of a square and two congruent rectangles. Its total area is $x^2 + 35x$ square units.

1. What is the length of the unlabeled side of each of the two rectangles?

2. If we add lines to make the figure a square, what is the area of the entire figure?

3. How is the process of finding the area of the entire figure like the process of building perfect squares for expressions like $x^2 + bx$?

NAME _____ DATE _____ PERIOD _____

 ## Summary
Completing the Square (Part 1)

Turning an expression into a perfect square can be a good way to solve a quadratic equation. Suppose we wanted to solve $x^2 - 14x + 10 = -30$.

The expression on the left, $x^2 - 14x + 10$, is not a perfect square, but $x^2 - 14x + 49$ *is* a perfect square. Let's transform that side of the equation into a perfect square (while keeping the equality of the two sides).

- One helpful way to start is by first moving the constant that is not a perfect square out of the way. Let's subtract 10 from each side:

$$x^2 - 14x + 10 - 10 = -30 - 10$$
$$x^2 - 14x = -40$$

- And then add 49 to each side:

$$x^2 - 14x + 49 = -40 + 49$$
$$x^2 - 14x + 49 = 9$$

- The left side is now a perfect square because it's equivalent to $(x - 7)(x - 7)$ or $(x - 7)^2$. Let's rewrite it:

$$(x - 7)^2 = 9$$

- If a number squared is 9, the number has to be 3 or -3. To finish up:

$$x - 7 = 3 \quad \text{or} \quad x - 7 = -3$$
$$x = 10 \quad \text{or} \quad x = 4$$

This method of solving quadratic equations is called **completing the square**. In general, perfect squares in standard form look like $x^2 + bx + \left(\frac{b}{2}\right)^2$, so to complete the square, take half of the coefficient of the linear term and square it.

In the example, half of -14 is -7, and $(-7)^2$ is 49. We wanted to make the left side $x^2 - 14x + 49$. To keep the equation true and maintain equality of the two sides of the equation, we added 49 to *each* side.

Glossary

completing the square

Practice
Completing the Square (Part 1)

1. Add the number that would make the expression a perfect square. Next, write an equivalent expression in factored form.

   a. $x^2 - 6x$

   b. $x^2 + 2x$

   c. $x^2 + 14x$

   d. $x^2 - 4x$

   e. $x^2 + 24x$

2. Mai is solving the equation $x^2 + 12x = 13$. She writes:

   $$x^2 + 12x = 13$$
   $$(x + 6)^2 = 49$$
   $$x = 1 \text{ or } x = \text{-}13$$

   Jada looks at Mai's work and is confused. She doesn't see how Mai got her answer.

   Complete Mai's missing steps to help Jada see how Mai solved the equation.

3. Match each equation to an equivalent equation with a perfect square on one side.

   A. $x^2 + 8x = 2$

   B. $x^2 + 10x = \text{-}13$

   C. $x^2 - 14x = 5$

   D. $x^2 + 2x = 0$

   E. $x^2 + 4x - 5 = 0$

   F. $x^2 - 20x = \text{-}9$

   1. $(x - 7)^2 = 54$

   2. $(x + 5)^2 = 12$

   3. $(x - 10)^2 = 91$

   4. $(x + 4)^2 = 18$

   5. $(x + 1)^2 = 1$

   6. $(x + 2)^2 = 9$

NAME _____ DATE _____ PERIOD _____

4. Solve each equation by completing the square.

$x^2 - 6x + 5 = 12$ $\qquad\qquad$ $x^2 - 2x = 8$

$11 = x^2 + 4x - 1$ $\qquad\qquad$ $x^2 - 18x + 60 = -21$

5. Rewrite each expression in standard form. **(Lesson 7-8)**

a. $(x + 3)(x - 3)$

b. $(7 + x)(x - 7)$

c. $(2x - 5)(2x + 5)$

d. $\left(x + \dfrac{1}{8}\right)\left(x - \dfrac{1}{8}\right)$

6. To find the product $203 \cdot 197$ without a calculator, Priya wrote $(200 + 3)$ $(200 - 3)$. Very quickly, and without writing anything else, she arrived at 39,991. Explain how writing the two factors as a sum and a difference may have helped Priya. **(Lesson 7-8)**

7. A basketball is dropped from the roof of a building and its height in feet is modeled by the function h. **(Lesson 6-5)**

Here is a graph representing h.

Select **all** the true statements about this situation.

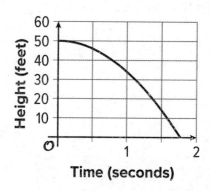

(A.) When $t = 0$ the height is 0 feet.

(B.) The basketball falls at a constant speed.

(C.) The expression that defines h is linear.

(D.) The expression that defines h is quadratic.

(E.) When $t = 0$ the ball is about 50 feet above the ground.

(F.) The basketball lands on the ground about 1.75 seconds after it is dropped.

8. A group of students are guessing the
   number of paper clips in a small box.

   (Lesson 4-13)

   The guesses and the guessing errors are
   plotted on a coordinate plane.

   What is the actual number of paper clips
   in the box?

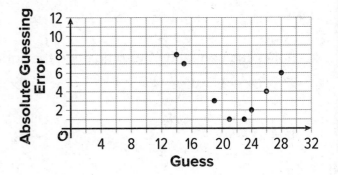

Lesson 7-13

Completing the Square (Part 2)

NAME _____ DATE _____ PERIOD _____

Learning Goal Let's solve some harder quadratic equations.

 Warm Up
13.1 Math Talk: Equations with Fractions

Solve each equation mentally.

$x + x = \dfrac{1}{4}$

$\left(\dfrac{3}{2}\right)^2 = x$

$\dfrac{3}{5} + x = \dfrac{9}{5}$

$\dfrac{1}{12} + x = \dfrac{1}{4}$

 Activity
13.2 Solving Some Harder Equations

Solve these equations by completing the square.

1. $(x - 3)(x + 1) = 5$

2. $x^2 + \dfrac{1}{2}x = \left(\dfrac{3}{16}\right)$

3. $x^2 + 3x + \dfrac{8}{4} = 0$

4. $(7 - x)(3 - x) + 3 = 0$

5. $x^2 + 1.6x + 0.63 = 0$

1. Show that the equation $x^2 + 10x + 9 = 0$ is equivalent to $(x + 3)^2 + 4x = 0$.

2. Write an equation that is equivalent to $x^2 + 9x + 16 = 0$ and that includes $(x + 4)^2$.

3. Does this method help you find solutions to the equations? Explain your reasoning.

NAME _____ DATE _____ PERIOD _____

Activity

13.3 Spot Those Errors!

Here are four equations, followed by worked solutions of the equations. Each solution has at least one error.

- Solve one or more of these equations by completing the square.

- Then, look at the worked solution of the same equation as the one you solved. Find and describe the error or errors in the worked solution.

1. $x^2 + 14x = -24$

2. $x^2 - 10x + 16 = 0$

3. $x^2 + 2.4x = -0.8$

4. $x^2 - \dfrac{6}{5}x + \dfrac{1}{5} = 0$

Worked solutions (with errors):

1.

$$x^2 + 14x = -24$$
$$x^2 + 14x + 28 = 4$$
$$(x + 7)^2 = 4$$

$$x + 7 = 2 \quad \text{or} \quad x + 7 = -2$$
$$x = -5 \quad \text{or} \quad x = -9$$

2.

$$x^2 - 10x + 16 = 0$$
$$x^2 - 10x + 25 = 9$$
$$(x - 5)^2 = 9$$

$$x - 5 = 9 \quad \text{or} \quad x - 5 = -9$$
$$x = 14 \quad \text{or} \quad x = -4$$

3.

$$x^2 + 2.4x = -0.8$$
$$x^2 + 2.4x + 1.44 = 0.64$$
$$(x + 1.2)^2 = 0.64$$
$$x + 1.2 = 0.8$$
$$x = -0.4$$

4.

$$x^2 - \frac{6}{5}x + \frac{1}{5} = 0$$
$$x^2 - \frac{6}{5}x + \frac{9}{25} = \frac{9}{25}$$
$$\left(x - \frac{3}{5}\right)^2 = \frac{9}{25}$$

$$x - \frac{3}{5} = \frac{3}{5} \quad \text{or} \quad x - \frac{3}{5} = -\frac{3}{5}$$
$$x = \frac{6}{5} \quad \text{or} \quad x = 0$$

NAME _____ DATE _____ PERIOD _____

Summary
Completing the Square (Part 2)

Completing the square can be a useful method for solving quadratic equations in cases in which it is not easy to rewrite an expression in factored form. For example, let's solve this equation:

$$x^2 + 5x - \frac{75}{4} = 0$$

First, we'll add $\frac{75}{4}$ to each side to make things easier on ourselves.

$$x^2 + 5x - \frac{75}{4} + \frac{75}{4} = 0 + \frac{75}{4}$$
$$x^2 + 5x = \frac{75}{4}$$

To complete the square, take $\frac{1}{2}$ of the coefficient of the linear term 5, which is $\frac{5}{2}$, and square it, which is $\frac{25}{4}$. Add this to each side:

$$x^2 + 5x + \frac{25}{4} = \frac{75}{4} + \frac{25}{4}$$
$$x^2 + 5x + \frac{25}{4} = \frac{100}{4}$$

Notice that $\frac{100}{4}$ is equal to 25 and rewrite it:

$$x^2 + 5x + \frac{25}{4} = 25$$

Since the left side is now a perfect square, let's rewrite it:

$$\left(x + \frac{5}{2}\right)^2 = 25$$

For this equation to be true, one of these equations must true:

$$x + \frac{5}{2} = 5 \quad \text{or} \quad x + \frac{5}{2} = \text{-}5$$

To finish up, we can subtract $\frac{5}{2}$ from each side of the equal sign in each equation.

$$x = 5 - \frac{5}{2} \quad \text{or} \quad x = \text{-}5 - \frac{5}{2}$$
$$x = \frac{5}{2} \quad \text{or} \quad x = \text{-}\frac{15}{2}$$
$$x = 2\frac{1}{2} \quad \text{or} \quad x = \text{-}7\frac{1}{2}$$

It takes some practice to become proficient at completing the square, but it makes it possible to solve many more equations than you could by methods you learned previously.

Practice
Completing the Square (Part 2)

1. Add the number that would make the expression a perfect square. Next, write an equivalent expression in factored form.

   a. $x^2 + 3x$

   b. $x^2 + 0.6x$

   c. $x^2 - 11x$

   d. $x^2 - \dfrac{5}{2}x$

   e. $x^2 + x$

2. Noah is solving the equation $x^2 + 8x + 15 = 3$. He begins by rewriting the expression on the left in factored form and writes $(x + 3)(x + 5) = 3$. He does not know what to do next.

   Noah knows that the solutions are $x = -2$ and $x = -6$, but is not sure how to get to these values from his equation.

   Solve the original equation by completing the square.

NAME _____ DATE _____ PERIOD _____

3. An equation and its solutions are given. Explain or show how to solve the equation by completing the square.

   a. $x^2 + 20x + 50 = 14$. The solutions are $x = -18$ and $x = -2$.

   b. $x^2 + 1.6x = 0.36$. The solutions are $x = -1.8$ and $x = 0.2$.

   c. $x^2 - 5x = \frac{11}{4}$. The solutions are $x = \frac{11}{2}$ and $x = \frac{-1}{2}$.

4. Solve each equation.

   a. $x^2 - 0.5x = 0.5$

   b. $x^2 + 0.8x = 0.09$

   c. $x^2 + \frac{13}{3}x = \frac{56}{36}$

5. Match each quadratic expression given in factored form with an equivalent expression in standard form. One expression in standard form has no match. (Lesson 7-8)

A. $(2 + x)(2 - x)$

B. $(x + 9)(x - 9)$

C. $(2 + x)(x - 2)$

D. $(x + y)(x - y)$

1. $x^2 - 4$

2. $81 - x^2$

3. $x^2 - y^2$

4. $4 - x^2$

5. $x^2 - 81$

6. Four students solved the equation $x^2 + 225 = 0$. Their work is shown here. Only one student solved it correctly. Determine which student solved the equation correctly. For each of the incorrect solutions, explain the mistake. (Lesson 7-9)

Student A:

$$x^2 + 225 = 0$$
$$x^2 = -225$$
$$x = 15 \quad \text{or} \quad x = -15$$

Student B:

$$x^2 + 225 = 0$$
$$x^2 = -225$$

No solutions

Student C:

$$x^2 + 225 = 0$$
$$(x - 15)(x + 15) = 0$$
$$x = 15 \quad \text{or} \quad x = -15$$

Student D:

$$x^2 + 225 = 0$$
$$x^2 = 225$$
$$x = 15 \quad \text{or} \quad x = -15$$

Lesson 7-14

Completing the Square (Part 3)

NAME _____ DATE _____ PERIOD _____

Learning Goal Let's complete the square for some more complicated expressions.

Warm Up
14.1 Perfect Squares in Two Forms

Elena says, "$(x + 3)^2$ can be expanded into $x^2 + 6x + 9$. Likewise, $(2x + 3)^2$ can be expanded into $4x^2 + 6x + 9$."

Find an error in Elena's statement and correct the error. Show your reasoning.

Activity
14.2 Perfect in A Different Way

1. Write each expression in standard form:

   a. $(4x + 1)^2$

   b. $(5x - 2)^2$

   c. $\left(\frac{1}{2}x + 7\right)^2$

   d. $(3x + n)^2$

   e. $(kx + m)^2$

2. Decide if each expression is a perfect square. If so, write an equivalent expression of the form $(kx + m)^2$. If not, suggest one change to turn it into a perfect square.

a. $4x^2 + 12x + 9$

b. $4x^2 + 8x + 25$

NAME _____ DATE _____ PERIOD _____

Activity
14.3 When All the Stars Align

1. Find the value of c to make each expression in the left column a perfect square in standard form. Then, write an equivalent expression in the form of squared factors. In the last row, write your own pair of equivalent expressions.

Standard Form ($ax^2 + bx + c$)	Squared Factors ($kx + m)^2$
$100x^2 + 80x + c$	
$36x^2 - 60x + c$	
$25x^2 + 40x + c$	
$0.25x^2 - 14x + c$	

2. Solve each equation by completing the square:

$$25x^2 + 40x = \text{-}12 \qquad\qquad 36x^2 - 60x + 10 = \text{-}6$$

Activity

14.4 Putting Stars into Alignment

Here are three methods for solving $3x^2 + 8x + 5 = 0$.

Try to make sense of each method.

Method 1:

$$3x^2 + 8x + 5 = 0$$
$$(3x + 5)(x + 1) = 0$$
$$x = -\frac{5}{3} \quad \text{or} \quad x = -1$$

Method 2:

$$3x^2 + 8x + 5 = 0$$
$$9x^2 + 24x + 15 = 0$$
$$(3x)^2 + 8(3x) + 15 = 0$$
$$U^2 + 8U + 15 = 0$$
$$(U + 5)(U + 3) = 0$$

$U = -5$ or $U = -3$

$3x = -5$ or $3x = -3$

$x = -\frac{5}{3}$ or $x = -1$

Method 3:

$$3x^2 + 8x + 5 = 0$$
$$9x^2 + 24x + 15 = 0$$
$$9x^2 + 24x + 16 = 0$$
$$(3x + 4)^2 = 0$$

$3x + 4 = 1$ or $3x + 4 = -1$

$x = -1$ or $x = -\frac{5}{3}$

Once you understand the methods, use each method at least one time to solve these equations.

1. $5x^2 + 17x + 6 = 0$

2. $6x^2 + 19x = -10$

3. $8x^2 - 33x + 4 = 0$

4. $8x^2 - 26x = -21$

5. $10x^2 + 37x = 36$

6. $12x^2 + 20x - 77 = 0$

NAME _____ DATE _____ PERIOD _____

Are you ready for more?

Find the solutions to $3x^2 - 6x + \dfrac{9}{4} = 0$. Explain your reasoning.

Summary
Completing the Square (Part 3)

In earlier lessons, we worked with perfect squares such as $(x + 1)^2$ and $(x - 5)$ $(x - 5)$. We learned that their equivalent expressions in standard form follow a predictable pattern:

- In general, $(x + m)^2$ can be written as $x^2 + 2mx + m^2$.

- If a quadratic expression is of the form $ax^2 + bx + c$ and is a perfect square, and the value of a is 1, then the value of b is $2m$, and the value of c is m^2 for some value of m.

In this lesson, the variable in the factors being squared had a coefficient other than 1, for example $(3x + 1)^2$ and $(2x - 5)(2x - 5)$. Their equivalent expression in standard form also followed the same pattern we saw earlier.

Squared Factors	Standard Form
$(3x + 1)^2$	$(3x)^2 + 2(3x)(1) + 1^2$ or $9x^2 + 6x + 1$
$(2x - 5)^2$	$(2x)^2 + 2(2x)(-5) + (-5)^2$ or $4x^2 - 20x + 25$

In general, $(kx + m)^2$ can be written as:

$$(kx)^2 + 2(kx)(m) + m^2 \qquad \text{or} \qquad k^2x^2 + 2kmx + m^2$$

If a quadratic expression is of the form $ax^2 + bx + c$, then:

- the value of a is k^2

- the value of b is $2km$

- the value of c is m^2

We can use this pattern to help us complete the square and solve equations when the squared term x^2 has a coefficient other than 1—for example: $16x^2 + 40x = 11$.

What constant term c can we add to make the expression on the left of the equal sign a perfect square? And how do we write this expression as squared factors?

- 16 is 4^2, so the squared factors could be $(4x + m)^2$.

- 40 is equal to $2(4m)$, so $2(4m) = 40$ or $8m = 40$. This means that $m = 5$.

- If c is m^2, then $c = 5^2$ or $c = 25$.

- So the expression $16x^2 + 40x + 25$ is a perfect square and is equivalent to $(4x + 5)^2$.

Let's solve the equation $16x^2 + 40x = 11$ by completing the square!

$$16x^2 + 40x = 11$$
$$16x^2 + 40x + 25 = 11 + 25$$
$$(4x + 5)^2 = 36$$

$$
\begin{array}{lll}
4x + 5 = 6 & \text{or} & 4x + 5 = \text{-}6 \\
4x = 1 & \text{or} & 4x = \text{-}11 \\
x = \dfrac{1}{4} & \text{or} & x = \text{-}\dfrac{11}{4}
\end{array}
$$

NAME _____ DATE _____ PERIOD _____

Practice
Completing the Square (Part 3)

1. Select **all** expressions that are perfect squares.

   (A.) $9x^2 + 24x + 16$

   (B.) $2x^2 + 20x + 100$

   (C.) $(7 - 3x)^2$

   (D.) $(5x + 4)(5x - 4)$

   (E.) $(1 - 2x)(-2x + 1)$

   (F.) $4x^2 + 6x + \frac{9}{4}$

2. Find the missing number that makes the expression a perfect square. Next, write the expression in factored form.

   a. $49x^2 - \underline{\hspace{1cm}}x + 16$

   b. $36x^2 + \underline{\hspace{1cm}}x + 4$

   c. $4x^2 - \underline{\hspace{1cm}}x + 25$

   d. $9x^2 + \underline{\hspace{1cm}}x + 9$

   e. $121x^2 + \underline{\hspace{1cm}}x + 9$

3. Find the missing number that makes the expression a perfect square. Next, write the expression in factored form.

   a. $9x^2 + 42x + \underline{\hspace{1cm}}$

   b. $49x^2 - 28x + \underline{\hspace{1cm}}$

   c. $25x^2 + 110x + \underline{\hspace{1cm}}$

   d. $64x^2 - 144x + \underline{\hspace{1cm}}$

   e. $4x^2 + 24x + \underline{\hspace{1cm}}$

4. Respond to each question.

a. Find the value of c to make the expression a perfect square. Then, write an equivalent expression in factored form.

Standard Form $ax^2 + bx + c$	Factored Form $(kx + m)^2$
$4x^2 + 4x$	
$25x^2 - 30x$	

b. Solve each equation by completing the square.

$4x^2 + 4x = 3$

$25x^2 - 30x + 8 = 0$

NAME _____ DATE _____ PERIOD _____

5. For each function *f*, decide if the equation $f(x) = 0$ has 0, 1, or 2
   solutions. Explain how you know. **(Lesson 7-5)**

A
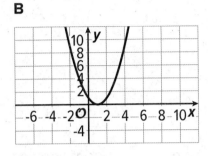

B
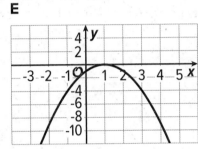

C
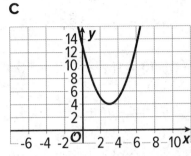

D
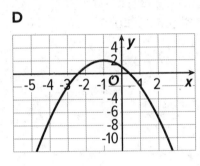

E

F
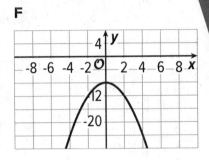

6. Solve each equation. **(Lesson 7-9)**

   $p^2 + 10 = 7p$ $x^2 + 11x + 27 = 3$ $(y + 2)(y + 6) = -3$

7. Which function could represent the height in meters of an object thrown upwards from a height of 25 meters above the ground t seconds after being launched? (Lesson 6-6)

Ⓐ $f(t) = -5t^2$

Ⓑ $f(t) = -5t^2 + 25$

Ⓒ $f(t) = -5t^2 + 25t + 50$

Ⓓ $f(t) = -5t^2 + 50t + 25$

8. A group of children are guessing the number of pebbles in a glass jar. The guesses and the guessing errors are plotted on a coordinate plane. (Lesson 4-13)

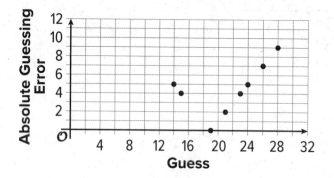

a. Which guess is furthest away from the actual number?

b. How far is the furthest guess away from the actual number?

Lesson 7-15

Quadratic Equations with Irrational Solutions

NAME _____ DATE _____ PERIOD _____

Learning Goal Let's find exact solutions to quadratic equations even if the solutions are irrational.

Warm Up
15.1 Roots of Squares

Here are some squares whose vertices are on a grid.

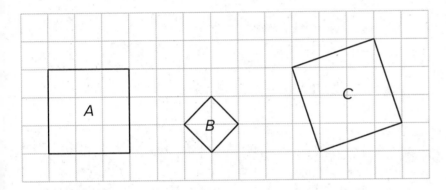

Find the area and the side length of each square.

Square	Area (Square Units)	Side Length (Units)
A		
B		
C		

Activity

15.2 Solutions Written as Square Roots

Solve each equation. Use the $\pm$ notation when appropriate.

1. $x^2 - 13 = -12$

2. $(x - 6)^2 = 0$

3. $x^2 + 9 = 0$

4. $x^2 = 18$

5. $x^2 + 1 = 18$

6. $(x + 1)^2 = 18$

NAME _____ DATE _____ PERIOD _____

Activity

15.3 Finding Irrational Solutions by Completing the Square

Here is an example of an equation being solved by graphing and by completing the square.

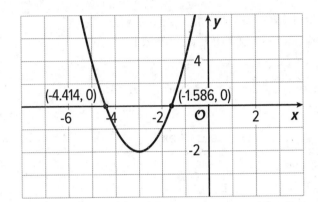

$x^2 + 6x + 7 = 0$

$x^2 + 6x + 9 = 2$

$(x + 3)^2 = 2$

$x + 3 = \pm\sqrt{2}$

$x = -3 \pm \sqrt{2}$

Verify: $\sqrt{2}$ is approximately 1.414.
So $-3 + \sqrt{2} \approx -1.586$ and $-3 - \sqrt{2} \approx -4.414$.

For each equation, find the exact solutions by completing the square and the approximate solutions by graphing. Then, verify that the solutions found using the two methods are close. If you get stuck, study the example.

1. $x^2 + 4x + 1 = 0$

2. $x^2 - 10x + 18 = 0$

3. $x^2 + 5x + \dfrac{1}{4} = 0$

4. $x^2 + \dfrac{8}{3}x + \dfrac{14}{9} = 0$

Are you ready for more?

Write a quadratic equation of the form $ax^2 + bx + c = 0$ whose solutions are $x = 5 - \sqrt{2}$ and $x = 5 + \sqrt{2}$.

When solving quadratic equations, it is important to remember that:

- Any positive number has two square roots, one positive and one negative, because there are two numbers that can be squared to make that number. (For example, 6^2 and $(-6)^2$ both equal 36, so 6 and -6 are both square roots of 36.)

- The square root symbol ($\sqrt{}$) can be used to express the *positive* square root of a number. For example, the square root of 36 is 6, but it can also be written as $\sqrt{36}$ because $\sqrt{36} \cdot \sqrt{36} = 36$.

- To express the negative square root of a number, say 36, we can write -6 or $-\sqrt{36}$.

- When a number is not a perfect square—for example, 40—we can express its square roots by writing $\sqrt{40}$ and $-\sqrt{40}$.

How could we write the solutions to an equation like $(x + 4)^2 = 11$? This equation is saying, "something squared is 11." To make the equation true, that something must be $\sqrt{11}$ or $-\sqrt{11}$. We can write:

$$x + 4 = \sqrt{11} \quad \text{or} \quad x + 4 = -\sqrt{11}$$
$$x = -4 + \sqrt{11} \quad \text{or} \quad x = -4 - \sqrt{11}$$

A more compact way to write the two solutions to the equation is:
$x = -4 \pm \sqrt{11}$.

About how large or small are those numbers? Are they positive or negative? We can use a calculator to compute the approximate values of both expressions:

$$-4 + \sqrt{11} \approx -0.683 \quad \text{or} \quad -4 - \sqrt{11} \approx -7.317$$

We can also approximate the solutions by graphing. The equation $(x + 4)^2 = 11$ is equivalent to $(x + 4)^2 - 11 = 0$, so we can graph the function $y = (x + 4)^2 - 11$ and find its zeros by locating the x-intercepts of the graph.

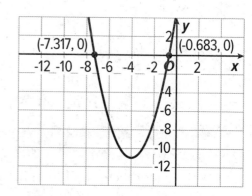

Glossary

irrational number

NAME _____ DATE _____ PERIOD _____

 Practice

Quadratic Equations with Irrational Solutions

1. Solve each equation and write the solutions using $\pm$ notation.

   a. $x^2 = 144$

   b. $x^2 = 5$

   c. $4x^2 = 28$

   d. $x^2 = \dfrac{25}{4}$

   e. $2x^2 = 22$

   f. $7x^2 = 16$

2. Match each expression to an equivalent expression.

   A. 4 ± 1

   B. $10 \pm \sqrt{4}$

   C. -6 ± 11

   D. $4 \pm \sqrt{10}$

   E. $\sqrt{16} \pm \sqrt{2}$

   1. -17 and 5

   2. $4 + \sqrt{2}$ and $4 - \sqrt{2}$

   3. 8 and 12

   4. 3 and 5

   5. $4 + \sqrt{10}$ and $4 - \sqrt{10}$

3. Respond to each question.

   a. Is $\sqrt{4}$ a positive or negative number? Explain your reasoning.

   b. Is $\sqrt{5}$ a positive or negative number? Explain your reasoning.

   c. Explain the difference between $\sqrt{9}$ and the solutions to $x^2 = 9$.

4. *Technology required.* For each equation, find the exact solutions by completing the square and the approximate solutions by graphing. Then, verify that the solutions found using the two methods are close.

$x^2 + 10x + 8 = 0$

$x^2 - 4x - 11 = 0$

5. Jada is working on solving a quadratic equation, as shown here.

$$p^2 - 5x = 0$$
$$p(p - 5) = 0$$
$$p - 5 = 0$$
$$p = 5$$

She thinks that her solution is correct because substituting 5 for p in the original expression $p^2 - 5p$ gives $5^2 - 5(5)$, which is $25 - 25$ or 0.

Explain the mistake that Jada made and show the correct solutions.

6. Which expression in factored form is equivalent to $30x^2 + 31x + 5$?

(Lesson 7-10)

(A.) $(6x + 5)(5x + 1)$

(B.) $(5x + 5)(6x + 1)$

(C.) $(10x + 5)(3x + 1)$

(D.) $(30x + 5)(x + 1)$

7. Two rocks are launched straight up in the air. The height of Rock A is given by the function f, where $f(t) = 4 + 30t - 16t^2$. The height of Rock B is given by g, where $g(t) = 5 + 20t - 16t^2$. In both functions, t is time measured in seconds after the rocks are launched and height is measured in feet above the ground. (Lesson 6-6)

a. Which rock is launched from a higher point?

b. Which rock is launched with a greater velocity?

8. Respond to each question. (Lesson 4-14)

   a. Describe how the graph of $f(x) = |x|$ has to be shifted to match the given graph.

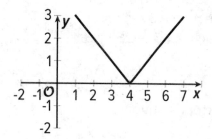

   b. Find an equation for the function represented by the graph.

Lesson 7-16

The Quadratic Formula

NAME _____ DATE _____ PERIOD _____

Learning Goal Let's learn a formula for finding solutions to quadratic equations.

Warm Up
16.1 Evaluate It

Each expression represents two numbers. Evaluate the expressions and find the two numbers.

1. $1 \pm \sqrt{49}$

2. $\dfrac{8 \pm 2}{5}$

3. $\pm\sqrt{(-5)^2 - 4 \cdot 4 \cdot 1}$

4. $\dfrac{-18 \pm \sqrt{36}}{2 \cdot 3}$

Activity
16.2 Pesky Equations

Choose one equation to solve, either by rewriting it in factored form or by completing the square. Be prepared to explain your choice of method.

1. $x^2 - 2x - 1.25 = 0$

2. $5x^2 + 9x - 44 = 0$

3. $x^2 + 1.25x = 0.375$

4. $4x^2 - 28x + 29 = 0$

Activity

16.3 Meet the Quadratic Formula

Here is a formula called the **quadratic formula**.

$$x = \frac{-b \pm \sqrt{b^2 - 4ac}}{2a}$$

The formula can be used to find the solutions to any quadratic equation in the form of $ax^2 + bx + c = 0$, where a, b, and c are numbers and a is not 0.

This example shows how it is used to solve $x^2 - 8x + 15 = 0$, in which $a = 1$, $b = -8$, and $c = 15$.

$$x = \frac{-b \pm \sqrt{b^2 - 4ac}}{2a}$$ original equation

$$x = \frac{-(-8) \pm \sqrt{(-8)^2 - 4(1)(15)}}{2(1)}$$ substitute the values of a, b, and c

$$x = \frac{8 \pm \sqrt{64 - 60}}{2}$$ evaluate each part of the expression

$$x = \frac{8 \pm \sqrt{4}}{2}$$

$$x = \frac{8 \pm 2}{2}$$

$$x = \frac{10}{2} \quad \text{or} \quad x = \frac{6}{2}$$

$$x = 5 \quad \text{or} \quad x = 3$$

NAME _____ DATE _____ PERIOD _____

Here are some quadratic equations and their solutions. Use the quadratic formula to show that the solutions are correct.

1. $x^2 + 4x - 5 = 0$. The solutions are $x = -5$ and $x = 1$.

2. $x^2 + 7x + 12 = 0$. The solutions are $x = -3$ and $x = -4$.

3. $x^2 + 10x + 18 = 0$. The solutions are $x = -5 \pm \frac{\sqrt{28}}{2}$.

4. $x^2 - 8x + 11 = 0$. The solutions are $x = 4 \pm \frac{\sqrt{20}}{2}$.

5. $9x^2 - 6x + 1 = 0$. The solution is $x = \frac{1}{3}$.

6. $6x^2 + 9x - 15 = 0$. The solutions are $x = -\frac{5}{2}$ and $x = 1$.

Are you ready for more?

1. Use the quadratic formula to solve $ax^2 + c = 0$. Let's call the resulting equation P.

2. Solve the equation $3x^2 - 27 = 0$ in two ways, showing your reasoning for each:

- Without using any formulas.

- Using equation P.

NAME _____ DATE _____ PERIOD _____

3. Check that you got the same solutions using each method.

4. Use the quadratic formula to solve $ax^2 + bx = 0$. Let's call the resulting equation Q.

5. Solve the equation $2x^2 + 5x = 0$ in two ways, showing your reasoning for each:

   • Without using any formulas.

   • Using equation Q.

6. Check that you got the same solutions using each method.

Summary
The Quadratic Formula

We have learned a couple of methods for solving quadratic equations algebraically:

- by rewriting the equation as factored form = 0 and using the zero product property

- by completing the square

Some equations can be solved quickly with one of these methods, but many cannot. Here is an example: $5x^2 - 3x - 1 = 0$. The expression on the left cannot be rewritten in factored form with rational coefficients. Because the coefficient of the squared term is not a perfect square, and the coefficient of the linear term is an odd number, completing the square would be inconvenient and would result in a perfect square with fractions.

The **quadratic formula** can be used to find the solutions to any quadratic equation, including those that are tricky to solve with other methods.

For an equation of the form $ax^2 + bx + c = 0$, where a, b, and c are numbers and $a \neq 0$, the solutions are given by:

$$x = \frac{-b \pm \sqrt{b^2 - 4ac}}{2a}$$

For the equation $5x^2 - 3x - 1 = 0$, we see that $a = 5$, $b = -3$, and $c = -1$. Let's solve it!

$$x = \frac{-b \pm \sqrt{b^2 - 4ac}}{2a} \qquad \text{the quadratic formula}$$

$$x = \frac{-(-3) \pm \sqrt{(-3)^2 - 4(5)(-1)}}{2(5)} \qquad \text{substitute the values of } a, b, \text{and } c$$

$$x = \frac{3 \pm \sqrt{9 + 20}}{10} \qquad \text{evaluate each part of the expression}$$

$$x = \frac{3 \pm \sqrt{29}}{10}$$

A calculator gives approximate solutions of 0.84 and -0.24 for $\frac{3 + \sqrt{29}}{10}$ and $\frac{3 - \sqrt{29}}{10}$.

We can also use the formula for simpler equations like $x^2 - 9x + 8 = 0$, but it may not be the most efficient way. If the quadratic expression can be easily rewritten in factored form or made into a perfect square, those methods may be preferable. For example, rewriting $x^2 - 9x + 8 = 0$ as $(x - 1)(x - 8) = 0$ immediately tells us that the solutions are 1 and 8.

Glossary

quadratic formula

NAME _____ DATE _____ PERIOD _____

Practice
The Quadratic Formula

1. For each equation, identify the values of a, b, and c that you would substitute into the quadratic formula to solve the equation.

   a. $3x^2 + 8x + 4 = 0$

   b. $2x^2 - 5x + 2 = 0$

   c. $-9x^2 + 13x - 1 = 0$

   d. $x^2 + x - 11 = 0$

   e. $-x^2 + 16x + 64 = 0$

2. Use the quadratic formula to show that the given solutions are correct.

   a. $x^2 + 9x + 20 = 0$. The solutions are $x = -4$ and $x = -5$.

   b. $x^2 - 10x + 21 = 0$. The solutions are $x = 3$ and $x = 7$.

   c. $3x^2 - 5x + 1 = 0$. The solutions are $x = \dfrac{5}{6} \pm \dfrac{\sqrt{13}}{6}$.

3. Select **all** the equations that are equivalent to $81x^2 + 180x - 200 = 100$.

   (Lesson 7-14)

   A. $81x^2 + 180x - 100 = 0$

   B. $81x^2 + 180x + 100 = 200$

   C. $81x^2 + 180x + 100 = 400$

   D. $(9x + 10)^2 = 400$

   E. $(9x + 10)^2 = 0$

   F. $(9x - 10)^2 = 10$

   G. $(9x - 10)^2 = 20$

4. *Technology required.* Two objects are launched upward. Each function gives the distance from the ground in meters as a function of time, t, in seconds.

Object A: $f(t) = 25 + 20t - 5t^2$ Object B: $g(t) = 30 + 10t - 5t^2$

Use graphing technology to graph each function. (Lesson 6-6)

   a. Which object reaches the ground first? Explain how you know.

   b. What is the maximum height of each object?

5. Identify the values of a, b, and c that you would substitute into the quadratic formula to solve the equation.

   a. $x^2 + 9x + 18 = 0$

   b. $4x^2 - 3x + 11 = 0$

   c. $81 - x + 5x^2 = 0$

   d. $\frac{4}{5}x^2 + 3x = \frac{1}{3}$

   e. $121 = x^2$

   f. $7x + 14x^2 = 42$

6. On the same coordinate plane, sketch a graph of each function. (Lesson 4-14)

- Function v, defined by $v(x) = |x + 6|$
- Function z, defined by $z(x) = |x| + 9$

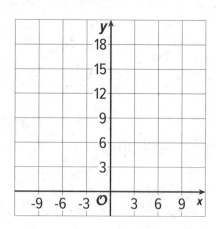

Lesson 7-17

Applying the Quadratic Formula (Part 1)

NAME _____ DATE _____ PERIOD _____

Learning Goal Let's use the quadratic formula to solve some problems.

Warm Up
17.1 No Solutions for You!

Here is an example of someone solving a quadratic equation that has no solutions:

$$(x + 3)^2 + 9 = 0$$
$$(x + 3)^2 = \text{-}9$$
$$x + 3 = \pm\sqrt{\text{-}9}$$

1. Study the example. At what point did you realize the equation had no solutions?

2. Explain how you know the equation $49 + x^2 = 0$ has no solutions.

Activity

17.2 The Potato and the Pumpkin

Answer each question without graphing. Explain or show your reasoning.

1. The equation $h(t) = -16t^2 + 80t + 64$ represented the height, in feet, of a potato t seconds after it has been launched.

   a. Write an equation that can be solved to find when the potato hits the ground. Then solve the equation.

   b. Write an equation that can be solved to find when the potato is 40 feet off the ground. Then solve the equation.

2. The equation $g(t) = 2 + 23.7t - 4.9t^2$ models the height, in meters, of a pumpkin t seconds after it has been launched from a catapult.

   a. Is the pumpkin still in the air 8 seconds later? Explain or show how you know.

   b. At what value of t does the pumpkin hit the ground? Show your reasoning.

NAME _____ DATE _____ PERIOD _____

Activity
17.3 Back to the Framer

1. In an earlier lesson, we tried to frame a picture that was 7 inches by 4 inches using an entire sheet of paper that was 4 inches by 2.5 inches. One equation we wrote was $(7 + 2x)(4 + 2x) = 38$.

   a. Explain or show what the equation $(7 + 2x)(4 + 2x) = 38$ tells us about the situation and what it would mean to solve it. Use the diagram, as needed.

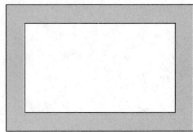

b. Solve the equation without graphing. Show your reasoning.

2. Suppose you have another picture that is 10 inches by 5 inches, and are now using a fancy paper that is 8.5 inches by 4 inches to frame the picture. Again, the frame is to be uniform in thickness all the way around. No fancy framing paper is to be wasted!

   Find out how thick the frame should be.

NAME _____ DATE _____ PERIOD _____

Are you ready for more?

Suppose that your border paper is 6 inches by 8 inches. You want to use all the paper to make a half-inch border around some rectangular picture.

1. Find two possible pairs of length and width of a rectangular picture that could be framed with a half-inch border and no leftover materials.

2. What must be true about the length and width of any rectangular picture that can be framed this way? Explain how you know.

Quadratic equations that represent situations cannot always be neatly put into factored form or easily solved by finding square roots. Completing the square is a workable strategy, but for some equations, it may involve many cumbersome steps. Graphing is also a handy way to solve the equations, but it doesn't always give us precise solutions.

With the quadratic formula, we can solve these equations more readily and precisely.

Here's an example: Function h models the height of an object, in meters, t seconds after it is launched into the air. It is is defined by $h(t) = -5t^2 + 25t$.

To know how much time it would take the object to reach 15 meters, we could solve the equation $15 = -5t^2 + 25t$. How should we do it?

- Rewriting it in standard form gives $-5t^2 + 25t - 15 = 0$. The expression on the left side of the equation cannot be written in factored form, however.

- Completing the square isn't convenient because the coefficient of the squared term is not a perfect square and the coefficient the linear term is an odd number.

- Let's use the quadratic formula, using $a = -5$, $b = 25$, and $c = -15$!

$$t = \frac{-b \pm \sqrt{b^2 - 4ac}}{2a}$$

$$t = \frac{-25 \pm \sqrt{25^2 - 4(-5)(-15)}}{2(-5)}$$

$$t = \frac{-25 \pm \sqrt{325}}{-10}$$

The expression $\dfrac{-25 \pm \sqrt{325}}{-10}$ represents the two exact solutions of the equation.

We can also get approximate solutions by using a calculator, or by reasoning that $\sqrt{325} \approx 18$.

The solutions tell us that there are two times after the launch when the object is at a height of 15 meters: at about 0.7 seconds (as the object is going up) and 4.3 seconds (as it comes back down).

NAME _____ DATE _____ PERIOD _____

Practice
Applying the Quadratic Formula (Part 1)

1. Select **all** the equations that have 2 solutions.

   (A.) $(x + 3)^2 = 9$

   (B.) $(x - 5)^2 = -5$

   (C.) $(x + 2)^2 - 6 = 0$

   (D.) $(x - 9)^2 + 25 = 0$

   (E.) $(x + 10)^2 = 1$

   (F.) $(x - 8)^2 = 0$

   (G.) $5 = (x + 1)(x + 1)$

2. A frog jumps in the air. The height, in inches, of the frog is modeled by the function $h(t) = 60t - 75t^2$, where t is the time after it jumped, measured in seconds.

   Solve $60t - 75t^2 = 0$. What do the solutions tell us about the jumping frog?

3. A tennis ball is hit straight up in the air, and its height, in feet above the ground, is modeled by the equation $f(t) = 4 + 12t - 16t^2$, where t is measured in seconds since the ball was thrown.

   a. Find the solutions to the equation $0 = 4 + 12t - 16t^2$.

   b. What do the solutions tell us about the tennis ball?

4. Rewrite each quadratic expression in standard form. (Lesson 7-10)

   a. $(x + 1)(7x + 2)$

   b. $(8x + 1)(x - 5)$

   c. $(2x + 1)(2x - 1)$

   d. $(4 + x)(3x - 2)$

5. Find the missing expression in parentheses so that each pair of quadratic expressions is equivalent. Show that your expression meets this requirement. (Lesson 7-10)

   a. $(4x - 1)(\underline{\hspace{2cm}})$ and $16x^2 - 8x + 1$

   b. $(9x + 2)(\underline{\hspace{2cm}})$ and $9x^2 - 16x - 4$

   c. $(\underline{\hspace{2cm}})(-x + 5)$ and $-7x^2 + 36x - 5$

NAME _____ DATE _____ PERIOD _____

6. The number of downloads of a song during a week is a function, f, of the number of weeks, w, since the song was released. The equation $f(w) = 100,000 \cdot \left(\frac{9}{10}\right)^w$ defines this function. **(Lesson 5-9)**

   a. What does the number 100,000 tell you about the downloads? What about the $\frac{9}{10}$?

   b. Is $f(-1)$ meaningful in this situation? Explain your reasoning.

7. Consider the equation $4x^2 - 4x - 15 = 0$. **(Lesson 7-16)**

   a. Identify the values of a, b, and c that you would substitute into the quadratic formula to solve the equation.

   b. Evaluate each expression using the values of a, b, and c.

   $-b$ b^2 $4ac$ $b^2 - 4ac$

   $\sqrt{b^2 - 4ac}$ $-b \pm \sqrt{b^2 - 4ac}$ $2a$ $\dfrac{-b \pm \sqrt{b^2 - 4ac}}{2a}$

   c. The solutions to the equation are $x = -\frac{3}{2}$ and $x = \frac{5}{2}$. Do these match the values of the last expression you evaluated in the previous question?

8. Respond to each question. **(Lesson 6-13)**

   a. Describe the graph of $y = -x^2$. (Does it open upward or downward? Where is its y-intercept? What about its x-intercepts?)

   b. Without graphing, describe how adding $16x$ to $-x^2$ would change each feature of the graph of $y = -x^2$. (If you get stuck, consider writing the expression in factored form.)

      i. the x-intercepts

      ii. the vertex

      iii. the y-intercept

      iv. the direction of opening of the U-shape graph

Lesson 7-18

Applying the Quadratic Formula (Part 2)

NAME _____ DATE _____ PERIOD _____

Learning Goal Let's use the quadratic formula and solve quadratic equations with care.

Warm Up
18.1 Bits and Pieces

Evaluate each expression for $a = 9$, $b = -5$, and $c = -2$

1. $-b$

2. b^2

3. $b^2 - 4ac$

4. $-b \pm \sqrt{a}$

Activity
18.2 Using the Formula with Care

Here are four equations, followed by attempts to solve them using the quadratic formula. Each attempt contains at least one error.

- Solve 1–2 equations by using the quadratic formula.

- Then, find and describe the error(s) in the worked solutions of the same equations as the ones you solved.

Equation 1: $2x^2 + 3 = 8x$ Equation 2: $x^2 + 3x = 10$

Equation 3: $9x^2 - 2x - 1 = 0$ Equation 4: $x^2 - 10x + 23 = 0$

Here are the worked solutions with errors:

Equation 1: $2x^2 + 3 = 8x$

$a = 2, b = -8, c = 3$

$$x = \frac{-b \pm \sqrt{b^2 - 4ac}}{2a}$$

$$x = \frac{-(-8) \pm \sqrt{(-8)^2 - 4(2)(3)}}{2(2)}$$

$$x = \frac{8 \pm \sqrt{64 - 24}}{4}$$

$$x = \frac{8 \pm \sqrt{40}}{4}$$

$$x = 2 \pm \sqrt{10}$$

Equation 2: $x^2 + 3x = 10$

$a = 1, b = 3, c = 10$

$$x = \frac{-b \pm \sqrt{b^2 - 4ac}}{2a}$$

$$x = \frac{-3 \pm \sqrt{3^2 - 4(1)(10)}}{2(1)}$$

$$x = \frac{-3 \pm \sqrt{9 - 40}}{2}$$

$$x = \frac{-3 \pm \sqrt{-31}}{2}$$

No solutions

Equation 3: $9x^2 - 2x - 1 = 0$

$a = 9, b = -2, c = -1$

$$x = \frac{-b \pm \sqrt{b^2 - 4ac}}{2a}$$

$$x = \frac{2 \pm \sqrt{(-2)^2 - 4(9)(-1)}}{2}$$

$$x = \frac{2 \pm \sqrt{4 + 36}}{2}$$

$$x = \frac{2 \pm \sqrt{40}}{2}$$

Equation 4: $x^2 - 10x + 23 = 0$

$a = 1, b = -10, c = 23$

$$x = \frac{-b \pm \sqrt{b^2 - 4ac}}{2a}$$

$$x = \frac{-10 \pm \sqrt{(-10)^2 - 4(1)(23)}}{2}$$

$$x = \frac{-10 \pm \sqrt{-100 - 92}}{2}$$

$$x = \frac{-10 \pm \sqrt{-192}}{2}$$

No solutions

NAME _____ DATE _____ PERIOD _____

Activity
18.3 Sure About That?

1. The equation $h(t) = 2 + 30t - 5t^2$ represents the height, as a function of time, of a pumpkin that was catapulted up in the air. Height is measured in meters and time is measured in seconds.

   a. The pumpkin reached a maximum height of 47 meters. How many seconds after launch did that happen? Show your reasoning.

   b. Suppose someone was unconvinced by your solution. Find another way (besides the steps you already took) to show your solution is correct.

2. The equation $r(p) = 80p - p^2$ models the revenue a band expects to collect as a function of the price of one concert ticket. Ticket prices and revenues are in dollars.

A band member says that a ticket price of either $15.50 or $74.50 would generate approximately $1,000 in revenue. Do you agree? Show your reasoning.

Are you ready for more?

Function g is defined by the equation $g(t) = 2 + 30t - 5t^2 - 47$. Its graph opens downward.

1. Find the zeros of function g without graphing. Show your reasoning.

2. Explain or show how the zeros you found can tell us the vertex of the graph of g.

3. Study the expressions that define functions g and h (which defined the height of the pumpkin). Explain how the maximum of function h, once we know it, can tell us the maximum of g.

NAME _____ DATE _____ PERIOD _____

Summary
Applying the Quadratic Formula (Part 2)

The quadratic formula has many parts in it. A small error in any one part can lead to incorrect solutions.

Suppose we are solving $2x^2 - 6 = 11x$. To use the formula, let's rewrite it in the form of $ax^2 + bx + c = 0$, which gives: $2x^2 - 11x - 6 = 0$.

Here are some common errors to avoid:

- Using the wrong values for a, b, and c in the formula.

$$x = \frac{-b \pm \sqrt{b^2 - 4ac}}{2a}$$

$$x = \frac{11 \pm \sqrt{(-11)^2 - 4(2)(-6)}}{2(2)}$$

$$x = \frac{-11 \pm \sqrt{(-11)^2 - 4(2)(-6)}}{2(2)}$$

That's better!

Nope! b is -11, so $-b$ is -(-11), which is 11, not -11.

- Forgetting to multiply 2 by a for the denominator in the formula.

$$x = \frac{11 \pm \sqrt{(-11)^2 - 4(2)(-6)}}{2}$$

$$x = \frac{11 \pm \sqrt{(-11)^2 - 4(2)(-6)}}{2(2)}$$

Nope! The denominator is $2a$, which is 2(2) or 4.

That's better!

- Forgetting that squaring a negative number produces a positive number.

$$x = \frac{11 \pm \sqrt{-121 - 4(2)(-6)}}{4}$$

$$x = \frac{11 \pm \sqrt{121 - 4(2)(-6)}}{4}$$

Nope! $(-11)^2$ is 121, not -121.

That's better!

- Forgetting that a negative number times a positive number is a negative number.

$$x = \frac{11 \pm \sqrt{121 - 48}}{4}$$

$$x = \frac{11 \pm \sqrt{121 + 48}}{4}$$

Nope! $4(2)(-6) = -48$ and $121 - (-48)$ is $121 + 48$.

That's better!

- Making calculation errors or not following the properties of algebra.

$$x = \frac{11 \pm \sqrt{169}}{4}$$

$$x = \frac{11 \pm 13}{4}$$

$$x = 11 \pm \sqrt{42.25}$$

That's better!

Nope! Both parts of the numerator, the 11 and the $\sqrt{169}$, get divided by 4. Also, $\frac{\sqrt{169}}{4}$ is not $\sqrt{42.25}$.

Let's finish by evaluating $\frac{11 \pm 13}{4}$ correctly:

$$x = \frac{11 + 13}{4} \quad \text{or} \quad x = \frac{11 - 13}{4}$$

$$x = \frac{24}{4} \quad \text{or} \quad x = -\frac{2}{4}$$

$$x = 6 \quad \text{or} \quad x = -\frac{1}{2}$$

To make sure our solutions are indeed correct, we can substitute the solutions back into the original equations and see whether each solution keeps the equation true.

Checking 6 as a solution:

$$2x^2 - 6 = 11x$$
$$2(6)^2 - 6 = 11(6)$$
$$2(36) - 6 = 66$$
$$72 - 6 = 66$$
$$66 = 66$$

Checking $-\frac{1}{2}$ as a solution:

$$2x^2 - 6 = 11x$$
$$2\left(-\frac{1}{2}\right)^2 - 6 = 11\left(-\frac{1}{2}\right)$$
$$2\left(\frac{1}{4}\right) - 6 = -\frac{11}{2}$$
$$\frac{1}{2} - 6 = -5\frac{1}{2}$$
$$-5\frac{1}{2} = -5\frac{1}{2}$$

We can also graph the equation $y = 2x^2 - 11x - 6$ and find its x-intercepts to see whether our solutions to $2x^2 - 11x - 6 = 0$ are accurate (or close to accurate).

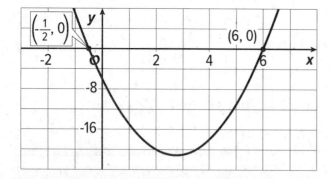

NAME _____ DATE _____ PERIOD _____

Practice
Applying the Quadratic Formula (Part 2)

1. Mai and Jada are solving the equation $2x^2 - 7x = 15$ using the quadratic formula but found different solutions.

Mai wrote:

$$x = \frac{-7 \pm \sqrt{7^2 - 4(2)(-15)}}{2(2)}$$

$$x = \frac{-7 \pm \sqrt{49 - (-120)}}{4}$$

$$x = \frac{-7 \pm \sqrt{169}}{4}$$

$$x = \frac{-7 \pm 13}{4}$$

$$x = -5 \quad \text{or} \quad x = \frac{3}{2}$$

Jada wrote:

$$x = \frac{-(-7) \pm \sqrt{-7^2 - 4(2)(-15)}}{2(2)}$$

$$x = \frac{7 \pm \sqrt{-49 - (-120)}}{4}$$

$$x = \frac{7 + \sqrt{71}}{4}$$

a. If this equation is written in standard form, $ax^2 + bx + c = 0$, what are the values of a, b, and c?

b. Do you agree with either of them? Explain your reasoning.

2. The equation $h(t) = -16t^2 + 80t + 64$ represents the height, in feet, of a potato t seconds after it was launched from a mechanical device.

a. Write an equation that would allow us to find the time the potato hits the ground.

b. Solve the equation without graphing. Show your reasoning.

3. Priya found $x = 3$ and $x = -1$ as solutions to $3x^2 - 6x - 9 = 0$. Is she correct? Show how you know.

4. Lin says she can tell that $25x^2 + 40x + 16$ and $49x^2 - 112x + 64$ are perfect squares because each expression has the following characteristics, which she saw in other perfect squares in standard form: (Lesson 7-11)

   • The first term is a perfect square. The last term is also a perfect square.

   • If we multiply a square root of the first term and a square root of the last term and then double the product, the result is the middle term.

   a. Show that each expression has the characteristics Lin described.

   b. Write each expression in factored form.

5. What are the solutions to the equation $2x^2 - 5x - 1 = 0$? (Lesson 7-16)

   (A.) $x = \dfrac{-5 \pm \sqrt{17}}{4}$

   (B.) $x = \dfrac{5 \pm \sqrt{17}}{4}$

   (C.) $x = \dfrac{-5 \pm \sqrt{33}}{4}$

   (D.) $x = \dfrac{5 \pm \sqrt{33}}{4}$

NAME _____ DATE _____ PERIOD _____

6. Solve each equation by rewriting the quadratic expression in factored form and using the zero product property, or by completing the square. Then, check if your solutions are correct by using the quadratic formula.

(Lesson 7-16)

a. $x^2 + 11x + 24 = 0$

b. $4x^2 + 20x + 25 = 0$

c. $x^2 + 8x = 5$

7. Here are the graphs of three equations.

   Match each graph with the appropriate
   equation. (Lesson 5-12)

   A. $y = 10\left(\dfrac{2}{3}\right)^x$ 1. X

   B. $y = 10\left(\dfrac{1}{4}\right)^x$ 2. Y

   C. $y = 10\left(\dfrac{3}{5}\right)^x$ 3. Z

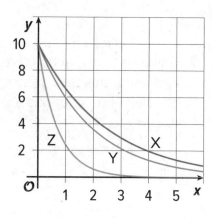

8. The function f is defined by $f(x) = (x + 1)(x + 6)$. (Lesson 6-11)

   a. What are the x-intercepts of the graph of f?

   b. Find the coordinates of the vertex of the graph of f. Show your
      reasoning.

   c. Sketch a graph of f.

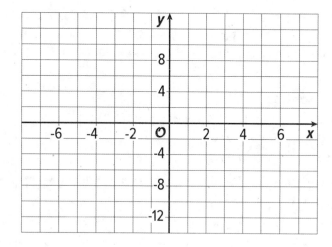

Lesson 7-19

Deriving the Quadratic Formula

NAME _____ DATE _____ PERIOD _____

Learning Goal Let's find out where the quadratic formula comes from.

 ## Warm Up
19.1 Studying Structure

Here are some perfect squares in factored and standard forms, and an expression showing how the two forms are related.

1. Study the first few examples, and then complete the missing numbers in the rest of the table.

Factored Form		Standard Form
$(x + 4)^2$	$(1x)^2 + 2(\ x)(\ ) + 4^2$	$x^2 + 8x + 16$
$(2x + 5)^2$	$(2x)^2 + 2(\ x)(\ ) + 5^2$	$4x^2 + 20x + 25$
$(3x - 4)^2$	$(3x)^2 + 2(\ x)(\ ) + (\ )^2$	$9x^2 - 24x + 16$
$(5x +\ )^2$	$(\ x)^2 + 2(\ x)(\ ) + (\ )^2$	$25x^2 + 30x +$
$(kx + m)^2$	$(\ x)^2 + 2(\ x)(\ ) + (\ )^2$	$x^2 +\quad x +$

2. Look at the expression in the last row of the table. If $ax^2 + bx + c$ is equivalent to $(kx + m)^2$, how are a, b, and c related to k and m?

Activity

19.2 Complete the Square using a Placeholder

1. One way to solve the quadratic equation $x^2 + 5x + 3 = 0$ is by completing the square. A partially solved equation is shown here. Study the steps.

   Then, knowing that P is a placeholder for $2x$, continue to solve for x but without evaluating any part of the expression. Be prepared to explain each step.

$x^2 + 5x + 3 = 0$	Original equation
$4x^2 + 20x + 12 = 0$	Multiply each side by 4
$4x^2 + 20x = \text{-}12$	Subtract 12 from each side
$(2x)^2 + 10(2x) = \text{-}12$	Rewrite $4x^2$ as $(2x)^2$ and $20x$ as $10(2x)$
$P^2 + 10P = \text{-}12$	Use P as a placeholder for $2x$

   $$P^2 + 10P + \underline{}^2 = \text{-}12 + \underline{}^2$$

   $$(P + \underline{})^2 = \text{-}12 + \underline{}^2$$

   $$P + \underline{} = \pm\sqrt{\text{-}12 + \underline{}^2}$$

   $$P = \underline{} \pm\sqrt{\text{-}12 + \underline{}^2}$$

   $$P = \underline{} \pm\sqrt{\underline{}^2 - 12}$$

   $$2x = \underline{} \pm\sqrt{\underline{}^2 - 12}$$

   $$x =$$

NAME _____ DATE _____ PERIOD _____

2. Explain how the solution is related to the quadratic formula.

 ## Activity

19.3 Decoding the Quadratic Formula

Here is one way to make sense of how the quadratic formula came about. Study the derivation until you can explain what happened in each step. Record your explanation next to each step.

$$ax^2 + bx + c = 0$$

$$4a^2x^2 + 4abx + 4ac = 0$$

$$4a^2x^2 + 4abx = -4ac$$

$$(2ax)^2 + 2b(2ax) = -4ac$$

$$M^2 + 2bM = -4ac$$

$$M^2 + 2bM + b^2 = -4ac + b^2$$

$$(M + b)^2 = b^2 - 4ac$$

$$M + b = \pm\sqrt{b^2 - 4ac}$$

$$M = -b \pm \sqrt{b^2 - 4ac}$$

$$2ax = -b \pm \sqrt{b^2 - 4ac}$$

$$x = \frac{-b \pm \sqrt{b^2 - 4ac}}{2a}$$

Here is another way to derive the quadratic formula by completing the square.

- First, divide each side of the equation $ax^2 + bx + c = 0$ by a to get $x^2 + \frac{b}{a}x + \frac{c}{a} = 0$.

- Then, complete the square for $x^2 + \frac{b}{a}x + \frac{c}{a} = 0$.

1. The beginning steps of this approach are shown here. Briefly explain what happens in each step.

$$x^2 + \frac{b}{a}x + \frac{c}{a} = 0 \qquad \text{Original equation}$$

$$x^2 + \frac{b}{a}x = -\frac{c}{a} \qquad \text{[1]}$$

$$x^2 + 2\left(\frac{b}{2a}\right)x + \left(\frac{b}{2a}\right)^2 = -\frac{c}{a} + \left(\frac{b}{2a}\right)^2 \qquad \text{[2]}$$

$$\left(x + \frac{b}{2a}\right)^2 = -\frac{c}{a} + \frac{b^2}{4a^2} \qquad \text{[3]}$$

$$\left(x + \frac{b}{2a}\right)^2 = -\frac{4ac}{4a^2} + \frac{b^2}{4a^2} \qquad \text{[4]}$$

$$\left(x + \frac{b}{2a}\right)^2 = \frac{b^2 - 4ac}{4a^2} \qquad \text{[5]}$$

$$x + \frac{b}{2a} = \pm\sqrt{\frac{b^2 - 4ac}{4a^2}} \qquad \text{[6]}$$

$$x + \frac{b}{2a} = \pm\frac{\sqrt{b^2 - 4ac}}{\sqrt{4a^2}} \qquad \text{[7]}$$

NAME _____ DATE _____ PERIOD _____

2. Continue the solving process until you have the equation
$x = \dfrac{-b \pm \sqrt{b^2 - 4ac}}{2a}$.

Summary
Deriving the Quadratic Formula

Recall that any quadratic equation can be solved by completing the square. The quadratic formula is essentially what we get when we put all the steps taken to complete the square for $ax^2 + bx + c = 0$ into a single expression.

When we expand a squared factor like $(3x + 5)^2$, the result is $(3x)^2 + 2(5)(3x) + 25$. Notice how the $(3x)$ appears in two places. If we replace $(3x)$ with another letter like P, we have $P^2 + 10P + 25$, which is a recognizable perfect square.

Likewise, if we expand $(kx + m)^2$, we have $(kx)^2 + 2m(kx) + m^2$. Replacing kx with P gives $P^2 + 2mP + m^2$, also a recognizable perfect square.

To complete the square is essentially to make one side of the equation have the same structure as $(kx)^2 + 2m(kx) + m^2$. Substituting a letter for (kx) makes it easier to see what is needed to complete the square. Let's complete the square for $ax^2 + bx + c = 0$!

- Start by subtracting c from each side.

$$ax^2 + bx = \text{-}c$$

- Next, let's multiply both sides by $4a$. On the left, this gives $4a^2$, a perfect square for the coefficient of x^2.

$$4a^2x^2 + 4abx = \text{-}4ac$$

- $4a^2x^2$ can be written $(2ax)^2$, and $4abx$ can be written $2b(2ax)$.

$$(2ax)^2 + 2b(2ax) = \text{-}4ac$$

- Let's replace $(2ax)$ with the letter P.

$$P^2 + 2bP = \text{-}4ac$$

- b^2 is the constant term that completes the square, so let's add b^2 to each side.

$$P^2 + 2bP + b^2 = \text{-}4ac + b^2$$

- The left side is now a perfect square and can be written as a squared factor.

$$(P + b)^2 = b^2 - 4ac$$

- The square roots of the expression on the right are the values of $P + b$.

$$P + b = \pm\sqrt{b^2 - 4ac}$$

- Once P is isolated, we can write $2ax$ in its place and solve for x.

$$P = \text{-}b \pm \sqrt{b^2 - 4ac}$$
$$2ax = \text{-}b \pm \sqrt{b^2 - 4ac}$$

- The solution is the quadratic formula!

$$x = \frac{\text{-}b \pm \sqrt{b^2 - 4ac}}{2a}$$

NAME _____ DATE _____ PERIOD _____

Practice

Deriving the Quadratic Formula

1. Respond to each question.

   a. The quadratic equation $x^2 + 7x + 10 = 0$ is in the form of $ax^2 + bx + c = 0$. What are the values of a, b, and c?

   b. Some steps for solving the equation by completing the square have been started here. In the third line, what might be a good reason for multiplying each side of the equation by 4?

$$x^2 + 7x + 10 = 0$$ Original equation

$$x^2 + 7x = -10$$ Subtract 10 from each side

$$4x^2 + 4(7x) = 4(-10)$$ Multiply each side by 4

$$(2x)^2 + 2(7)2x + \_^2 = \_^2 - 4(10)$$ Rewrite $4x^2$ as $(2x)^2$ and $4(7x)$ as $2(7)2x$

$$(2x + \_)^2 = \_^2 - 4(10)$$

$$2x + \_ = \pm\sqrt{\_^2 - 4(10)}$$

$$2x = \_ \pm \sqrt{\_^2 - 4(10)}$$

$$x =$$

   c. Complete the unfinished steps, and explain what happens in each step in the second half of the solution.

   d. Substitute the values of a, b, and c into the quadratic formula, $x = \dfrac{-b \pm \sqrt{b^2 - 4ac}}{2a}$, but do not evaluate any of the expressions. Explain how this expression is related to solving $x^2 + 7x + 10 = 0$ by completing the square.

2. Consider the equation $x^2 - 39 = 0$.

   a. Does the quadratic formula work to solve this equation? Explain or show how you know.

   b. Can you solve this equation using square roots? Explain or show how you know.

3. Clare is deriving the quadratic formula by solving $ax^2 + bx + c = 0$ by completing the square.

   She arrived at this equation. $(2ax + b)^2 = b^2 - 4ac$

   Briefly describe what she needs to do to finish solving for x and then show the steps.

4. Tyler is solving the quadratic equation $x^2 + 8x + 11 = 4$.

   (Lesson 7-12)

   Study his work and explain the mistake he made. Then, solve the equation correctly.

   $$x^2 + 8x + 11 = 4$$
   $$x^2 + 8x + 16 = 4$$
   $$(x + 4)^2 = 4$$
   $$x = -8 \qquad \text{or} \qquad x = 0$$

NAME _____ DATE _____ PERIOD _____

5. Solve the equation by using the quadratic formula. Then, check if your solutions are correct by rewriting the quadratic expression in factored form and using the zero product property. (Lesson 7-16)

a. $2x^2 - 3x - 5 = 0$

b. $x^2 - 4x = 21$

c. $3 - x - 4x^2 = 0$

6. A tennis ball is hit straight up in the air, and its height, in feet above the ground, is modeled by the equation $f(t) = 4 + 12t - 16t^2$, where t is measured in seconds since the ball was thrown. (Lesson 7-17)

a. Find the solutions to $6 = 4 + 12t - 16t^2$ without graphing. Show your reasoning.

b. What do the solutions say about the tennis ball?

7. Consider the equation $y = 2x(6 - x)$. (Lesson 6-11)

   a. What are the x-intercepts of the graph of this equation? Explain how you know.

   b. What is the x-coordinate of the vertex of the graph of this equation? Explain how you know.

   c. What is the y-coordinate of the vertex? Show your reasoning.

   d. Sketch the graph of this equation.

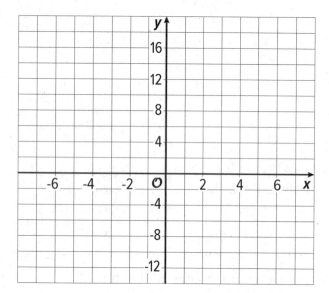

Lesson 7-20

Rational and Irrational Solutions

NAME _____ DATE _____ PERIOD _____

Learning Goal Let's consider the kinds of numbers we get when solving quadratic equations.

Warm Up
20.1 Rational or Irrational?

Numbers like -1.7, $\sqrt{16}$, and $\frac{5}{3}$ are known as *rational numbers.*

Numbers like $\sqrt{12}$ and $\sqrt{\frac{5}{9}}$ are known as *irrational numbers.*

Here is a list of numbers. Sort them into rational and irrational.

97   -8.2 $\sqrt{5}$ $-\frac{3}{7}$ $\sqrt{100}$ $\sqrt{\frac{9}{4}}$ $-\sqrt{18}$

Activity

20.2 Suspected Irrational Solutions

1. Graph each quadratic equation using graphing technology. Identify the zeros of the function that the graph represents, and say whether you think they might be rational or irrational. Be prepared to explain your reasoning.

Equations	Zeros	Rational or Irrational?
$y = x^2 - 8$		
$y = (x - 5)^2 - 1$		
$y = (x - 7)^2 - 2$		
$y = \left(\frac{x}{4}\right)^2 - 5$		

2. Find exact solutions (not approximate solutions) to each equation and show your reasoning. Then, say whether you think each solution is rational or irrational. Be prepared to explain your reasoning.

   a. $x^2 - 8 = 0$

   b. $(x - 5)^2 = 1$

   c. $(x - 7)^2 = 2$

   d. $\left(\frac{x}{4}\right)^2 - 5 = 0$

NAME _____ DATE _____ PERIOD _____

Activity

20.3 Experimenting with Rational and Irrational Numbers

Here is a list of numbers:

2   3 $\frac{1}{3}$ 0 $\sqrt{2}$ $\sqrt{3}$ $-\sqrt{3}$ $\frac{1}{\sqrt{3}}$

Here are some statements about the sums and products of numbers. For each statement, decide whether it is *always* true, true for *some* numbers but not others, or *never* true.

1. Sums:

   a. The sum of two rational numbers is rational.

   b. The sum of a rational number and an irrational number is irrational.

   c. The sum of two irrational numbers is irrational.

2. Products:

   a. The product of two rational numbers is rational.

   b. The product of a rational number and an irrational number is irrational.

   c. The product of two irrational numbers is irrational.

Experiment with sums and products of two numbers in the given list to help you decide.

It can be quite difficult to show that a number is irrational. To do so, we have to explain why the number is impossible to write as a ratio of two integers. It took mathematicians thousands of years before they were finally able to show that π is irrational, and they still don't know whether or not π^π is irrational.

Here is a way we could show that $\sqrt{2}$ can't be rational, and is therefore irrational.

- Let's assume that $\sqrt{2}$ were rational and could be written as a fraction $\frac{a}{b}$, where a and b are non-zero integers.

- Let's also assume that a and b are integers that no longer have any common factors. For example, to express 0.4 as $\frac{a}{b}$, we write $\frac{2}{5}$ instead of $\frac{4}{10}$ or $\frac{200}{500}$. That is, we assume that a and b are 2 and 5, rather than 4 and 10, or 200 and 500.

1. If $\sqrt{2} = \frac{a}{b}$, then $2 = \dfrac{\Box}{\Box}$.

2. Explain why a^2 must be an even number.

3. Explain why if a^2 is an even number, then a itself is also an even number. (If you get stuck, consider squaring a few different integers.)

4. Because a is an even number, then a is 2 times another integer, say, k. We can write $a = 2k$. Substitute $2k$ for a in the equation you wrote in the first question. Then, solve for b^2.

5. Explain why the resulting equation shows that b^2, and therefore b, are also even numbers.

NAME _____ DATE _____ PERIOD _____

6. We just arrived at the conclusion that a and b are even numbers, but given our assumption about a and b, it is impossible for this to be true. Explain why this is.

If a and b cannot both be even, $\sqrt{2}$ must be equal to some number other than $\frac{a}{b}$.

Because our original assumption that we could write $\sqrt{2}$ as a fraction $\frac{a}{b}$ led to a false conclusion, that assumption must be wrong. In other words, we must not be able to write $\sqrt{2}$ as a fraction. This means $\sqrt{2}$ is irrational!

The solutions to quadratic equations can be rational or irrational. Recall that:

- *Rational numbers* are fractions and their opposites. Numbers like 12, -3, $\frac{5}{3}$, $\sqrt{25}$, -4.79, and $\sqrt{\frac{9}{16}}$ are rational. ($\sqrt{25}$ is a fraction, because it's equal to $\frac{5}{1}$. The number -4.79 is the opposite of 4.79, which is $\frac{479}{100}$.)

- Any number that is not rational is *irrational*. Some examples are $\sqrt{2}$, π, $-\sqrt{5}$, and $\sqrt{\frac{7}{2}}$. When an irrational number is written as a decimal, its digits do not stop or repeat, so a decimal can only approximate the value of the number.

How do we know if the solutions to a quadratic equation are rational or irrational?

If we solve a quadratic equation $ax^2 + bx + c = 0$ by graphing a corresponding function ($y = ax^2 + bx + c$), sometimes we can tell from the x-coordinates of the x-intercepts. Other times, we can't be sure.

Let's solve $x^2 - \frac{49}{100} = 0$ and $x^2 - 5 = 0$ by graphing $y = x^2 - \frac{49}{100}$ and $y = x^2 - 5$.

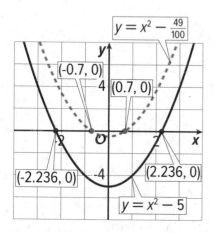

The graph of $y = x^2 - \frac{49}{100}$ crosses the x-axis at -0.7 and 0.7. There are no digits after the 7, suggesting that the x-values are exactly $-\frac{7}{10}$ and $\frac{7}{10}$, which are rational.

To verify that these numbers are exact solutions to the equation, we can see if they make the original equation true.

$(0.7)^2 - \frac{49}{100} = 0$ and $(-0.7)^2 - \frac{49}{100} = 0$, so ± 0.7 are exact solutions.

The graph of $y = x^2 - 5$, created using graphing technology, is shown to cross the x-axis at -2.236 and 2.236. It is unclear if the x-coordinates stop at three decimal places or if they continue. If they stop or eventually make a repeating pattern, the solutions would be rational. If they never stop or make a repeating pattern, the solutions would be irrational.

NAME _____ DATE _____ PERIOD _____

We can tell, though, that 2.236 is not an exact solution to the equation. Substituting 2.236 for x in the original equation gives $2.236^2 - 5$, which we can tell is close to 0 but is not exactly 0. This means ± 2.236 are not exact solutions, and the solutions may be irrational.

To be certain whether the solutions are rational or irrational, we can solve the equations.

- The solutions to $x^2 - \dfrac{49}{100} = 0$ are ± 0.7, which are rational.
- The solutions to $x^2 - 5 = 0$ are $\pm\sqrt{5}$, which are irrational. (2.236 is an approximation of $\sqrt{5}$, not equal to $\sqrt{5}$.)

What about a solution like $-4 + \sqrt{6}$, which is a sum of a rational number and an irrational one? Or a solution like $\dfrac{1}{5}\sqrt{3}$, which is a product of a rational number and an irrational number? Are they rational or irrational?

We will investigate solutions that are sums and products of different types of numbers in an upcoming lesson.

Practice

Rational and Irrational Solutions

1. Decide whether each number is rational or irrational.

   10 $\dfrac{4}{5}$ $\sqrt{4}$ $\sqrt{10}$ -3 $\sqrt{\dfrac{25}{4}}$ $\sqrt{0.6}$

2. Here are the solutions to some quadratic equations. Select **all** solutions that are rational.

   (A.) 5 ± 2

   (B.) $\sqrt{4} \pm 1$

   (C.) $\dfrac{1}{2} \pm 3$

   (D.) $10 \pm \sqrt{3}$

   (E.) $\pm\sqrt{25}$

   (F.) $1 \pm \sqrt{2}$

3. Solve each equation. Then, determine if the solutions are rational or irrational.

   a. $(x + 1)^2 = 4$

   b. $(x - 5)^2 = 36$

   c. $(x + 3)^2 = 11$

   d. $(x - 4)^2 = 6$

4. Here is a graph of the equation $y = 81(x - 3)^2 - 4$.

   a. Based on the graph, what are the solutions to the equation $81(x - 3)^2 = 4$?

   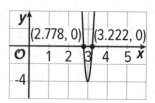

   b. Can you tell whether they are rational or irrational? Explain how you know.

NAME _____ DATE _____ PERIOD _____

c. Solve the equation using a different method and say whether the solutions are rational or irrational. Explain or show your reasoning.

5. Match each equation to an equivalent equation with a perfect square on one side. (Lesson 7-13)

A. $x^2 - 9x = \dfrac{1}{2}$

B. $x^2 + 6.4x - 8.9 = 0$

C. $x^2 - 5x = 11$

D. $x^2 + 0.1x + 0.0005 = 0$

E. $x^2 - \dfrac{6}{7}x = \dfrac{1}{49}$

F. $x^2 + 1.21x = 6.28$

1. $(x - 2.5)^2 = 17.25$

2. $\left(x - \dfrac{9}{2}\right)^2 = \dfrac{83}{4}$

3. $\left(x - \dfrac{3}{7}\right)^2 = \dfrac{10}{49}$

4. $(x + 0.05)^2 = 0.002$

5. $(x + 3.2)^2 = 19.14$

6. $(x + 0.605)^2 = 6.646025$

6. To derive the quadratic formula, we can multiply $ax^2 + bx + c = 0$ by an expression so that the coefficient of x^2 is a perfect square and the coefficient of x is an even number. (Lesson 7-19)

a. Which expression, a, $2a$, or $4a$, would you multiply $ax^2 + bx + c = 0$ by to get started deriving the quadratic formula?

b. What does the equation $ax^2 + bx + c = 0$ look like when you multiply both sides by your answer?

7. Here is a graph that represents $y = x^2$.

   On the same coordinate plane, sketch and label the graph that represents each equation: (Lesson 6-12)

   a. $y = -x^2 - 4$

   b. $y = 2x^2 + 4$

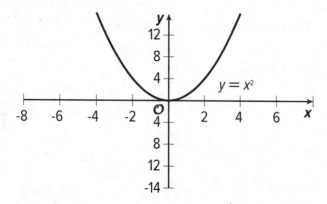

8. Which quadratic expression is in vertex form? (Lesson 6-15)

   Ⓐ $x^2 - 6x + 8$

   Ⓑ $(x - 6)^2 + 3$

   Ⓒ $(x - 3)(x - 6)$

   Ⓓ $(8 - x)x$

9. Function f is defined by the expression $\dfrac{5}{x - 2}$. (Lesson 4-10)

   a. Evaluate $f(12)$.

   b. Explain why $f(2)$ is undefined.

   c. Give a possible domain for f.

Lesson 7-21

Sums and Products of Rational and Irrational Numbers

NAME _____ DATE _____ PERIOD _____

Learning Goal Let's make convincing arguments about why the sums and products of rational and irrational numbers are always certain kinds of numbers.

Warm Up
21.1 Operations on Integers

Here are some examples of integers:

$$-25 \quad -10 \quad -2 \quad -1 \quad 0 \quad 5 \quad 9 \quad 40$$

1. Experiment with adding any two numbers from the list (or other integers of your choice). Try to find one or more examples of two integers that:

   a. add up to another integer

   b. add up to a number that is *not* an integer

2. Experiment with multiplying any two numbers from the list (or other integers of your choice). Try to find one or more examples of two integers that:

   a. multiply to make another integer

   b. multiply to make a number that is *not* an integer

1. Here are a few examples of adding two rational numbers. Is each sum a rational number? Be prepared to explain how you know.

   a. $4 + 0.175 = 4.175$

   b. $\dfrac{1}{2} + \dfrac{4}{5} = \dfrac{5}{10} + \dfrac{8}{10} = \dfrac{13}{10}$

   c. $-0.75 + \dfrac{14}{8} = \dfrac{-6}{8} + \dfrac{14}{8} = \dfrac{8}{8} = 1$

   d. a is an integer: $\dfrac{2}{3} + \dfrac{a}{15} = \dfrac{10}{15} + \dfrac{a}{15} = \dfrac{10 + a}{15}$

2. Here is a way to explain why the sum of two rational numbers is rational. Suppose $\dfrac{a}{b}$ and $\dfrac{c}{d}$ are fractions. That means that a, b, c, and d are integers, and b and d are not 0.

   a. Find the sum of $\dfrac{a}{b}$ and $\dfrac{c}{d}$. Show your reasoning.

   b. In the sum, are the numerator and the denominator integers? How do you know?

   c. Use your responses to explain why the sum of $\frac{a}{b} + \frac{c}{d}$ is a rational number.

3. Use the same reasoning as in the previous question to explain why the product of two rational numbers, $\frac{a}{b} \cdot \frac{c}{d}$, must be rational.

Are you ready for more?

Consider numbers that are of the form $a + b\sqrt{5}$, where a and b are integers. Let's call such numbers *quintegers*.

Here are some examples of quintegers:

- $3 + 4\sqrt{5}$ ($a = 3, b = 4$)
- $7 - 2\sqrt{5}$ ($a = 7, b = -2$)
- $-5 + \sqrt{5}$ ($a = -5, b = 1$)
- 3 ($a = 3, b = 0$).

1. When we add two quintegers, will we always get another quinteger? Either prove this, or find two quintegers whose sum is not a quinteger.

2. When we multiply two quintegers, will we always get another quinteger? Either prove this, or find two quintegers whose product is not a quinteger.

Activity

21.3 Sums and Products of Rational and Irrational Numbers

1. Here is a way to explain why $\sqrt{2} + \frac{1}{9}$ is irrational.

   - Let s be the sum of $\sqrt{2}$ and $\frac{1}{9}$, or $s = \sqrt{2} + \frac{1}{9}$.

   - Suppose s is rational.

   a. Would $s + -\frac{1}{9}$ be rational or irrational? Explain how you know.

   b. Evaluate $s + -\frac{1}{9}$. Is the sum rational or irrational?

   c. Use your responses so far to explain why s cannot be a rational number, and therefore $\sqrt{2} + \frac{1}{9}$ cannot be rational.

2. Use the same reasoning as in the earlier question to explain why $\sqrt{2} \cdot \frac{1}{9}$ is irrational.

NAME _____ DATE _____ PERIOD _____

Activity
21.4 Equations with Different Kinds of Solutions

1. Consider the equation $4x^2 + bx + 9 = 0$. Find a value of b so that the equation has:

   a. 2 rational solutions

   b. 2 irrational solutions

   c. 1 solution

   d. no solutions

2. Describe all the values of b that produce 2, 1, and no solutions.

3. Write a new quadratic equation with each type of solution. Be prepared to explain how you know that your equation has the specified type and number of solutions.

   a. no solutions

   b. 2 irrational solutions

   c. 2 rational solutions

   d. 1 solution

Summary
Sums and Products of Rational and Irrational Numbers

We know that quadratic equations can have rational solutions or irrational solutions. For example, the solutions to $(x + 3)(x - 1) = 0$ are -3 and 1, which are rational. The solutions to $x^2 - 8 = 0$ are $\pm\sqrt{8}$, which are irrational.

Sometimes solutions to equations combine two numbers by addition or multiplication—for example, $\pm 4\sqrt{3}$ and $1 + \sqrt{12}$. What kind of number are these expressions?

When we add or multiply two rational numbers, is the result rational or irrational?

- The sum of two rational numbers is rational. Here is one way to explain why it is true:

  - Any two rational numbers can be written $\frac{a}{b}$ and $\frac{c}{d}$, where a, b, c, and d are integers, and b and d are not zero.

  - The sum of $\frac{a}{b}$ and $\frac{c}{d}$ is $\frac{ad + bc}{bd}$. The denominator is not zero because neither b nor d is zero.

  - Multiplying or adding two integers always gives an integer, so we know that ad, bc, bd and $ad + bc$ are all integers.

  - If the numerator and denominator of $\frac{ad + bc}{bd}$ are integers, then the number is a fraction, which is rational.

- The product of two rational numbers is rational. We can show why in a similar way:

  - For any two rational numbers $\frac{a}{b}$ and $\frac{c}{d}$, where a, b, c, and d are integers, and b and d are not zero, the product is $\frac{ac}{bd}$.

  - Multiplying two integers always results in an integer, so both ac and bd are integers, so $\frac{ac}{bd}$ is a rational number.

What about two irrational numbers?

- The sum of two irrational numbers could be either rational or irrational. We can show this through examples:

  - $\sqrt{3}$ and $-\sqrt{3}$ are each irrational, but their sum is 0, which is rational.

  - $\sqrt{3}$ and $\sqrt{5}$ are each irrational, and their sum is irrational.

NAME _____ DATE _____ PERIOD _____

- The product of two irrational numbers could be either rational or irrational. We can show this through examples:

  - $\sqrt{2}$ and $\sqrt{8}$ are each irrational, but their product is $\sqrt{16}$ or 4, which is rational.

  - $\sqrt{2}$ and $\sqrt{7}$ are each irrational, and their product is $\sqrt{14}$, which is not a perfect square and is therefore irrational.

What about a rational number and an irrational number?

- The sum of a rational number and an irrational number is irrational. To explain why requires a slightly different argument:

  - Let R be a rational number and I an irrational number. We want to show that $R + I$ is irrational.

  - Suppose s represents the sum of R and I ($s = R + I$) and suppose s is rational.

  - If s is rational, then $s + -R$ would also be rational, because the sum of two rational numbers is rational.

  - $s + -R$ is not rational, however, because $(R + I) + -R = I$.

  - $s + -R$ cannot be both rational and irrational, which means that our original assumption that s was rational was incorrect. s, which is the sum of a rational number and an irrational number, must be irrational.

- The product of a non-zero rational number and an irrational number is irrational. We can show why this is true in a similar way:

  - Let R be rational and I irrational. We want to show that $R \cdot I$ is irrational.

  - Suppose p is the product of R and I ($p = R \cdot I$) and suppose p is rational.

  - If p is rational, then $p \cdot \frac{1}{R}$ would also be rational because the product of two rational numbers is rational.

  - $p \cdot \frac{1}{R}$ is not rational, however, because $R \cdot I \cdot \frac{1}{R} = I$.

  - $p \cdot \frac{1}{R}$ cannot be both rational and irrational, which means our original assumption that p was rational was false. p, which is the product of a rational number and an irrational number, must be irrational.

Practice

Sums and Products of Rational and Irrational Numbers

Match each expression to an equivalent expression. (Lesson 7-15)

A. $\sqrt{5} \pm \sqrt{3}$

B. $1 \pm \sqrt{3}$

C. $\sqrt{3} \pm 1$

D. 5 ± -2

E. -3 ± -3

1. 3 and 7

2. $\sqrt{5} + \sqrt{3}$ and $\sqrt{5} - \sqrt{3}$

3. -6 and 0

4. $\sqrt{3} + 1$ and $\sqrt{3} - 1$

5. $1 + \sqrt{3}$ and $1 - \sqrt{3}$

2. Consider the statement: "An irrational number multiplied by an irrational number always makes an irrational product."

   Select **all** the examples that show that this statement is false.

   Ⓐ $\sqrt{4} \cdot \sqrt{5}$

   Ⓑ $\sqrt{4} \cdot \sqrt{4}$

   Ⓒ $\sqrt{7} \cdot \sqrt{7}$

   Ⓓ $\frac{1}{\sqrt{5}} \cdot \sqrt{5}$

   Ⓔ $\sqrt{0} \cdot \sqrt{7}$

   Ⓕ $-\sqrt{5} \cdot \sqrt{5}$

   Ⓖ $\sqrt{5} \cdot \sqrt{7}$

3. Respond to each question. (Lesson 6-15)

   a. Where is the vertex of the graph that represents $y = (x - 3)^2 + 5$?

   b. Does the graph open up or down? Explain how you know.

4. Here are the solutions to some quadratic equations. Decide if the solutions are rational or irrational.

   $3 \pm \sqrt{2}$ $\sqrt{9} \pm 1$ $\frac{1}{2} \pm \frac{3}{2}$ 10 ± 0.3

   $\frac{1 \pm \sqrt{8}}{2}$ $-7 \pm \sqrt{\frac{4}{9}}$

NAME _____ DATE _____ PERIOD _____

5. Find an example that shows that the statement is false.

   a. An irrational number multiplied by an irrational number always makes an irrational product.

   b. A rational number multiplied by an irrational number never gives a rational product.

   c. Adding an irrational number to an irrational number always gives an irrational sum.

6. Which equation is equivalent to $x^2 - 3x = \frac{7}{4}$ but has a perfect square on one side? **(Lesson 7-13)**

   (A.) $x^2 - 3x + 3 = \frac{19}{4}$ **(C.)** $x^2 - 3x + \frac{9}{4} = \frac{16}{4}$

   (B.) $x^2 - 3x + \frac{3}{4} = \frac{10}{4}$ **(D.)** $x^2 - 3x + \frac{9}{4} = \frac{7}{4}$

7. A student who used the quadratic formula to solve $2x^2 - 8x = 2$ said that the solutions are $x = 2 + \sqrt{5}$ and $x = 2 - \sqrt{5}$. **(Lesson 7-18)**

   a. What equations can we graph to check those solutions? What features of the graph do we analyze?

   b. How do we look for $2 + \sqrt{5}$ and $2 - \sqrt{5}$ on a graph?

8. Here are 4 graphs. Match each graph with a quadratic equation that it represents. (Lesson 6-15)

Graph A

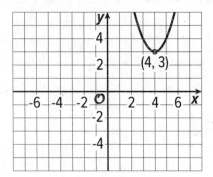

Graph B

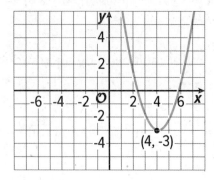

Graph C

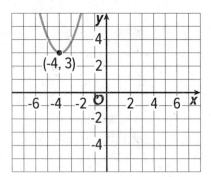

Graph D

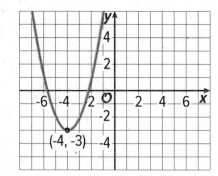

A. Graph A

B. Graph B

C. Graph C

D. Graph D

1. $y = (x + 4)^2 - 3$

2. $y = (x - 4)^2 - 3$

3. $y = (x + 4)^2 + 3$

4. $y = (x - 4)^2 + 3$

Lesson 7-22

Rewriting Quadratic Expressions in Vertex Form

NAME _____ DATE _____ PERIOD _____

Learning Goal Let's see what else completing the square can help us do.

Warm Up
22.1 Three Expressions, One Function

These expressions each define the same function.

$$x^2 + 6x + 8 \qquad (x + 2)(x + 4) \qquad (x + 3)^2 - 1$$

Without graphing or doing any calculations, determine where the following features would be on a graph that represents the function.

1. the vertex

2. the x-intercepts

3. the y-intercept

Activity
22.2 Back and Forth

1. Here are two expressions in vertex form. Rewrite each expression in standard form. Show your reasoning.

   a. $(x + 5)^2 + 1$

   b. $(x - 3)^2 - 7$

2. Think about the steps you took, and about reversing them. Try converting one or both of the expressions in standard form back into vertex form. Explain how you go about converting the expressions.

3. Test your strategy by rewriting $x^2 + 10x + 9$ in vertex form.

4. Let's check the expression you rewrote in vertex form.

   a. Use graphing technology to graph both $x^2 + 10x + 9$ and your new expression. Does it appear that they define the same function?

   b. If you convert your expression in vertex form back into standard form, do you get $x^2 + 10x + 9$?

 ## Activity

22.3 Inconvenient Coefficients

1. Respond to each question.

   a. Here is one way to rewrite $3x^2 + 12x + 9$ in vertex form. Study the steps and write a brief explanation of what is happening at each step.

   $3x^2 + 12x + 9$ Original expression

   $3(x^2 + 4x + 3)$

   $3(x^2 + 4x + 3 + 1 - 1)$

   $3(x^2 + 4x + 4 - 1)$

   $3((x + 2)^2 - 1)$

   $3(x + 2)^2 - 3$

NAME _____ DATE _____ PERIOD _____

b. What is the vertex of the graph that represents this expression?

c. Does the graph open upward or downward? Explain how you know.

2. Rewrite each expression in vertex form. Show your reasoning.

   a. $-2x^2 - 4x + 6$

   b. $4x^2 + 24x + 20$

   c. $-x^2 + 20x$

Are you ready for more?

1. Write $f(x) = 2(x - 3)(x - 9)$ in vertex form without completing the square. (Hint: Think about finding the zeros of the function.) Explain your reasoning.

2. Write $g(x) = 2(x - 3)(x - 9) + 21$ in vertex form without completing the square. Explain your reasoning.

Activity

22.4 Info Gap: Features of Functions

Your teacher will give you either a problem card or a data card. Do not show or read your card to your partner.

If your teacher gives you the data card:

1. Silently read the information on your card.

2. Ask your partner "What specific information do you need?" and wait for your partner to ask for information. Only give information that is on your card. (Do not figure out anything for your partner!)

3. Before telling your partner the information, ask "Why do you need to know (that piece of information)?"

4. Read the problem card, and solve the problem independently.

5. Share the data card, and discuss your reasoning.

If your teacher gives you the problem card:

1. Silently read your card and think about what information you need to answer the question.

2. Ask your partner for the specific information that you need.

3. Explain to your partner how you are using the information to solve the problem.

4. When you have enough information, share the problem card with your partner, and solve the problem independently.

5. Read the data card, and discuss your reasoning.

Pause here so your teacher can review your work. Ask your teacher for a new set of cards and repeat the activity, trading roles with your partner.

NAME _____ DATE _____ PERIOD _____

Summary
Rewriting Quadratic Expressions in Vertex Form

Remember that a quadratic function can be defined by equivalent expressions in different forms, which enable us to see different features of its graph. For example, these expressions define the same function:

$(x - 3)(x - 7)$ factored form

$x^2 - 10x + 21$ standard form

$(x - 5)^2 - 4$ vertex form

- From factored form, we can tell that the x-intercepts are (3, 0) and (7, 0).

- From standard form, we can tell that the y-intercept is (0, 21).

- From *vertex form*, we can tell that the vertex is (5, -4).

Recall that a function expressed in vertex form is written as: $a(x - h)^2 + k$. The values of h and k reveal the vertex of the graph: (h, k) are the coordinates of the vertex. In this example, a is 1, h is 5, and k is -4.

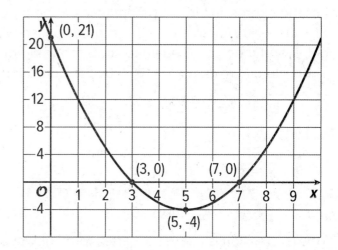

- If we have an expression in vertex form, we can rewrite it in standard form by using the distributive property and combining like terms.

  Let's say we want to rewrite $(x - 1)^2 - 4$ in standard form.

$$(x - 1)^2 - 4$$
$$(x - 1)(x - 1) - 4$$
$$x^2 - 2x + 1 - 4$$
$$x^2 - 2x - 3$$

- If we have an expression in standard form, we can rewrite it in vertex form by completing the square.

  Let's rewrite $x^2 + 10x + 24$ in vertex form.

A perfect square would be $x^2 + 10x + 25$, so we need to add 1. Adding 1, however, would change the expression. To keep the new expression equivalent to the original one, we will need to both add 1 and subtract 1.

$$x^2 + 10x + 24$$

$$x^2 + 10x + 24 + 1 - 1$$

$$x^2 + 10x + 25 - 1$$

$$(x + 5)^2 - 1$$

- Let's rewrite another expression in vertex form: $-2x^2 + 12x - 30$.

To make it easier to complete the square, we can use the distributive property to rewrite the expression with -2 as a factor, which gives $-2(x^2 - 6x + 15)$.

For the expression in the parentheses to be a perfect square, we need $x^2 - 6x + 9$. We have 15 in the expression, so we can subtract 6 from it to get 9, and then add 6 again to keep the value of the expression unchanged. Then, we can rewrite $x^2 - 6x + 9$ in factored form.

$$-2x^2 + 12x - 30$$

$$-2(x^2 - 6x + 15)$$

$$-2(x^2 - 6x + 15 - 6 + 6)$$

$$-2(x^2 - 6x + 9 + 6)$$

$$-2((x - 3)^2 + 6)$$

This expression is not yet in vertex form, however. To finish up, we need to apply the distributive property again so that the expression is of the form $a(x - h)^2 + k$:

$$-2((x - 3)^2 + 6)$$

$$-2(x - 3)^2 - 12$$

When written in this form, we can see that the vertex of the graph representing $-2(x - 3)^2 - 12$ is (3, -12).

Glossary

vertex form (of a quadratic expression)

NAME _____ DATE _____ PERIOD _____

Practice
Rewriting Quadratic Expressions in Vertex Form

1. The following quadratic expressions all define the same function.

$(x + 5)(x + 3)$ $x^2 + 8x + 15$ $(x + 4)^2 - 1$

Select **all** of the statements that are true about the graph of this function.

(A.) The y-intercept is (0, -15).

(B.) The vertex is (-4, -1).

(C.) The x-intercepts are (-5, 0) and (-3, 0).

(D.) The x-intercepts are (0, 5) and (0, 3).

(E.) The x-intercept is (0, 15).

(F.) The y-intercept is (0, 15).

(G.) The vertex is (4, -1).

2. The following expressions all define the same quadratic function.

$(x - 4)(x + 6)$ $x^2 + 2x - 24$ $(x + 1)^2 - 25$

   a. What is the y-intercept of the graph of the function?

   b. What are the x-intercepts of the graph?

   c. What is the vertex of the graph?

   d. Sketch a graph of the function without graphing technology. Make sure the x-intercepts, y-intercept, and vertex are plotted accurately.

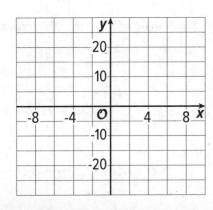

3. Here is one way an expression in standard form is rewritten into vertex form.

$x^2 - 7x + 6$	original expression
$x^2 - 7x + \left(-\dfrac{7}{2}\right)^2 + 6 - \left(-\dfrac{7}{2}\right)^2$	step 1
$\left(x - \dfrac{7}{2}\right)^2 + 6 - \dfrac{49}{4}$	step 2
$\left(x - \dfrac{7}{2}\right)^2 + \dfrac{24}{4} - \dfrac{49}{4}$	step 3
$\left(x - \dfrac{7}{2}\right)^2 - \dfrac{25}{4}$	step 4

a. In step 1, where did the number $-\dfrac{7}{2}$ come from?

b. In step 1, why was $\left(-\dfrac{7}{2}\right)^2$ added and then subtracted?

c. What happened in step 2?

d. What happened in step 3?

e. What does the last expression tell us about the graph of a function defined by this expression?

4. Rewrite each quadratic expression in vertex form.

a. $d(x) = x^2 + 12x + 36$

b. $f(x) = x^2 + 10x + 21$

c. $g(x) = 2x^2 - 20x + 32$

NAME _____ DATE _____ PERIOD _____

5. Respond to each question. **(Lesson 7-21)**

   a. Give an example that shows that the sum of two irrational numbers can be rational.

   b. Give an example that shows that the sum of two irrational numbers can be irrational.

6. Respond to each question. **(Lesson 7-21)**

   a. Give an example that shows that the product of two irrational numbers can be rational.

   b. Give an example that shows that the product of two irrational numbers can be irrational.

7. Select **all** the equations with irrational solutions. **(Lesson 7-15)**

   (A.) $36 = x^2$

   (B.) $x^2 = \frac{1}{4}$

   (C.) $x^2 = 8$

   (D.) $2x^2 = 8$

   (E.) $x^2 = 0$

   (F.) $x^2 = 40$

   (G.) $9 = x^2 - 1$

8. Respond to each question. (Lesson 6-16)

   a. What are the coordinates of the vertex of the graph of the function defined by $f(x) = 2(x + 1)^2 - 4$?

   b. Find the coordinates of two other points on the graph.

   c. Sketch the graph of f.

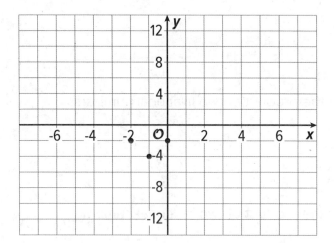

9. How is the graph of the equation $y = (x - 1)^2 + 4$ related to the graph of the equation $y = x^2$? (Lesson 6-17)

   Ⓐ The graph of $y = (x - 1)^2 + 4$ is the same as the graph of $y = x^2$ but is shifted 1 unit to the right and 4 units up.

   Ⓑ The graph of $y = (x - 1)^2 + 4$ is the same as the graph of $y = x^2$ but is shifted 1 unit to the left and 4 units up.

   Ⓒ The graph of $y = (x - 1)^2 + 4$ is the same as the graph of $y = x^2$ but is shifted 1 unit to the right and 4 units down.

   Ⓓ The graph of $y = (x - 1)^2 + 4$ is the same as the graph of $y = x^2$ but is shifted 1 unit to the left and 4 units down.

Lesson 7-23

Using Quadratic Expressions in Vertex Form to Solve Problems

NAME _____ DATE _____ PERIOD _____

Learning Goal Let's find the maximum or minimum value of a quadratic function.

Warm Up
23.1 Values of a Function

Here are graphs that represent two functions, f and g, defined by:

$f(x) = (x - 4)^2 + 1$

$g(x) = -(x - 12)^2 + 7$

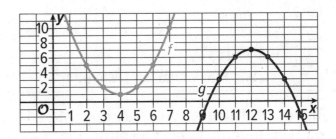

1. $f(1)$ can be expressed in words as "the value of f when x is 1." Find or compute:

   a. the value of f when x is 1

   b. $f(3)$

   c. $f(10)$

2. Can you find an x value that would make $f(x)$:

   a. Less than 1?

   b. Greater than 10,000?

3. $g(9)$ can be expressed in words as "the value of g when x is 9." Find or compute:

   a. the value of g when x is 9

   c. $g(2)$

   b. $g(13)$

4. Can you find an x value that would make $g(x)$:

   a. Greater than 7?

   b. Less than -10,000?

Activity

23.2 Maximums and Minimums

1. The graph that represents $p(x) = (x - 8)^2 + 1$ has its vertex at $(8, 1)$. Here is one way to show, without graphing, that $(8, 1)$ corresponds to the *minimum* value of p.

   - When $x = 8$, the value of $(x - 8)^2$ is 0, because $(8 - 8)^2 = 0^2 = 0$.

   - Squaring any number always results in a positive number, so when x is any value other than 8, $(x - 8)$ will be a number other than 0, and when squared, $(x - 8)^2$ will be positive.

   - Any positive number is greater than 0, so when $x \neq 8$, the value of $(x - 8)^2$ will be greater than when $x = 8$. In other words, p has the least value when $x = 8$.

   Use similar reasoning to explain why the point $(4, 1)$ corresponds to the *maximum* value of q, defined by $q(x) = -2(x - 4)^2 + 1$.

NAME _____ DATE _____ PERIOD _____

2. Here are some quadratic functions, and the coordinates of the vertex of the graph of each. Determine if the vertex corresponds to the maximum or the minimum value of the function. Be prepared to explain how you know.

Equation	Coordinates of The Vertex	Maximum or Minimum?
$f(x) = -(x - 4)^2 + 6$	(4, 6)	
$g(x) = (x + 7)^2 - 3$	(-7, -3)	
$h(x) = 4(x + 5)^2 + 7$	(-5, 7)	
$k(x) = x^2 - 6x - 3$	(3, -12)	
$m(x) = -x^2 + 8x$	(4, 16)	

Are you ready for more?

Here is a portion of the graph of function q, defined by $q(x) = -x^2 + 14x - 40$.

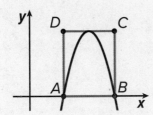

ABCD is a rectangle. Points A and B coincide with the x-intercepts of the graph, and segment CD just touches the vertex of the graph.

Find the area of *ABCD*. Show your reasoning.

A function A, defined by $p(600 - 75p)$, describes the revenue collected from the sales of tickets for Performance A, a musical.

The graph represents a function B that models the revenue collected from the sales of tickets for Performance B, a Shakespearean comedy.

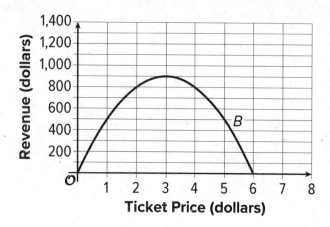

In both functions, p represents the price of one ticket, and both revenues and prices are measured in dollars.

Without creating a graph of A, determine which performance gives the greater maximum revenue when tickets are p dollars each. Explain or show your reasoning.

NAME _____ DATE _____ PERIOD _____

Summary
Using Quadratic Expressions in Vertex Form to Solve Problems

Any quadratic function has either a *maximum* or a *minimum* value. We can tell whether a quadratic function has a maximum or a minimum by observing the vertex of its graph.

Here are graphs representing functions f and g, defined by $f(x) = -(x + 5)^2 + 4$ and $g(x) = x^2 + 6x - 1$.

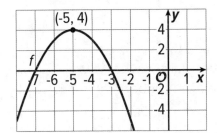

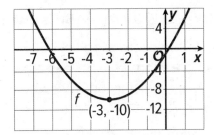

- The vertex of the graph of f is (-5, 4) and the graph is a U shape that opens downward.

- No other points on the graph of f (no matter how much we zoom out) are higher than (-5, 4), so we can say that f has a maximum of 4, and that this occurs when $x = -5$.

- The vertex of the graph of g is at (-3, -10) and the graph is a U shape that opens upward.

- No other points on the graph (no matter how much we zoom out) are lower than (-3, -10), so we can say that g has a minimum of -10, and that this occurs when $x = -3$.

We know that a quadratic expression in vertex form can reveal the vertex of the graph, so we don't actually have to graph the expression. But how do we know, without graphing, if the vertex corresponds to a maximum or a minimum value of a function?

The vertex form can give us that information as well!

To see if (-3,-10) is a minimum or maximum of g, we can rewrite $x^2 + 6x - 1$ in vertex form, which is $(x + 3)^2 - 10$. Let's look at the squared term in $(x + 3)^2 - 10$.

- When $x = -3$, $(x + 3)$ is 0, so $(x + 3)^2$ is also 0.

- When x is not -3, the expression $(x + 3)$ will be a non-zero number, and $(x + 3)^2$ will be positive (squaring any number gives a positive result).

- Because a squared number cannot have a value less than 0, $(x + 3)^2$ has the least value when $x = -3$.

To see if (-5, 4) is a minimum or maximum of f, let's look at the squared term in $-(x + 5)^2 + 4$.

- When $x = -5$, $(x + 5)$ is 0, so $(x + 5)^2$ is also 0.

- When x is not -5, the expression $(x + 5)$ will be non-zero, so $(x + 5)^2$ will be positive. The expression $-(x + 5)^2$ has a negative coefficient of -1, however. Multiplying $(x + 5)^2$ (which is positive when $x \neq -5$) by a negative number results in a negative number.

- Because a negative number is always less than 0, the value of $-(x + 5)^2 + 4$ will always be less when $x \neq -5$ than when $x = -5$. This means $x = -5$ gives the greatest value of f.

Glossary

maximum

minimum

NAME _____ DATE _____ PERIOD _____

Practice
Using Quadratic Expressions in Vertex Form to Solve Problems

1. Here is a graph of a quadratic function $f(x)$. What is the minimum value of $f(x)$?

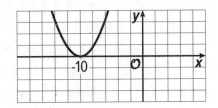

2. The graph that represents $f(x) = (x + 1)^2 - 4$ has its vertex at (-1, -4).

   Explain how we can tell from the expression $(x + 1)^2 - 4$ that -4 is the minimum value of f rather than the maximum value.

3. Each expression here defines a quadratic function. Find the vertex of the graph of the function. Then, state whether the vertex corresponds to the maximum or the minimum value of the function.

   a. $(x - 5)^2 + 6$

   b. $(x + 5)^2 - 1$

   c. $-2(x + 3)^2 - 10$

   d. $3(x - 7)^2 + 11$

   e. $-(x - 2)^2 - 2$

   f. $(x + 1)^2$

4. Consider the equation $x^2 = 12x$. (Lesson 7-19)

   a. Can we use the quadratic formula to solve this equation? Explain or show how you know.

   b. Is it easier to solve this equation by completing the square or by rewriting it in factored form and using the zero product property? Explain or show your reasoning.

5. Match each equation to the number of solutions it has. (Lesson 7-17)

   A. $(x - 1)(x - 5) = 5$ 1. no solutions

   B. $x^2 - 2x = -1$ 2. 1 solution

   C. $(x - 5)^2 = -25$ 3. 2 solutions

6. Which equation has irrational solutions? (Lesson 7-20)

   (A.) $100x^2 = 9$

   (B.) $9(x - 1)^2 = 4$

   (C.) $4x^2 - 1 = 0$

   (D.) $9(x + 3)^2 = 27$

NAME _____ DATE _____ PERIOD _____

7. Let I represent an irrational number and let R represent a rational number. Decide if each statement is true or false. Explain your thinking. (Lesson 7-21)

a. $R \cdot I$ can be rational.

b. $I \cdot I$ can be rational.

c. $R \cdot R$ can be rational.

8. Here are graphs of the equations $y = x^2$, $y = (x - 3)^2$, and $y = (x - 3)^2 + 7$. (Lesson 6-17)

a. How do the 3 graphs compare?

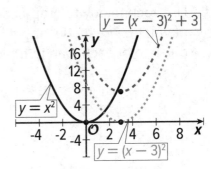

b. How does the -3 in $(x - 3)^2$ affect the graph?

c. How does the +7 in $(x - 3)^2 + 7$ affect the graph?

9. Three $5,000 loans have different annual interest rates. Loan A charges 10.5% annual interest, Loan B charges 15.75%, and Loan C charges 18.25%. (Lesson 5-15)

   a. If we graph the amount owed as a function of years without payment, what would the three graphs look like? Describe or sketch your prediction.

   b. Use technology to graph each function. Based on your graphs, if no payments are made, about how many years will it take for the unpaid balance of each loan to triple?

Lesson 7-24

Using Quadratic Equations to Model Situations and Solve Problems

NAME _____ DATE _____ PERIOD _____

Learning Goal Let's analyze a situation modeled by a quadratic equation.

 ## Warm Up
24.1 Equations of Two Lines and A Curve

1. Write an equation representing the line that passes through each pair of points.

   a. (3, 3) and (5, 5)

   b. (0, 4) and (-4, 0)

2. Solve this equation: $x + 1 = (x - 2)^2 - 3$. Show your reasoning.

 ## Activity
24.2 The Dive

The function h, defined by $h(t) = -5t^2 + 10t + 7.5$, models the height of a diver above the water (in meters), t seconds after the diver leaves the board. For each question, explain how you know.

1. How high above the water is the diving board?

2. When does the diver hit the water?

3. At what point during her descent toward the water is the diver at the same height as the diving board?

4. When does the diver reach the maximum height of the dive?

5. What is the maximum height the diver reaches during the dive?

NAME _____ DATE _____ PERIOD _____

Are you ready for more?

Another diver jumps off a platform, rather than a springboard. The platform is also 7.5 meters above the water, but this diver hits the water after about 1.5 seconds.

Write an equation that would approximately model her height over the water, h, in meters, t seconds after she has left the platform. Include the term $-5t^2$, which accounts for the effect of gravity.

 ## Activity

24.3 A Linear Function and A Quadratic Function

Here are graphs of a linear function and a quadratic function. The quadratic function is defined by the expression $(x - 4)^2 - 5$.

Find the coordinates of P, Q, and R without using graphing technology. Show your reasoning.

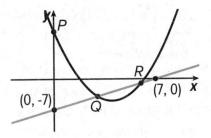

Certain real-world situations can be modeled by quadratic functions, and these functions can be represented by equations. Sometimes, all the skills we have developed are needed to make sense of these situations. When we have a mathematical model and the skills to use the model to answer questions, we are able to gain useful or interesting insights about the situation.

Suppose we have a model for the height of a launched object, h, as a function of time since it was launched t, defined by $h(t) = -4.9t^2 + 28t + 2.1$. We can answer questions such as these about the object's flight:

- From what height was the object launched?

  (An expression in standard form can help us with this question. Or, we can evaluate $h(0)$ to find the answer.)

- At what time did it hit the ground?

  (When an object hits the ground, its height is 0, so we can find the zeros using one of the methods we learned: graphing, rewriting in factored form, completing the square, or using the quadratic formula.)

- What was its maximum height, and at what time did it reach the maximum height?

  (We can rewrite the expression in vertex form, but we can also use the zeros of the function or a graph to do so.)

Sometimes, relationships between quantities can be effectively communicated with graphs and expressions rather than with words. For example, these graphs represent a linear function, f, and a quadratic function, g, with the same variables for their input and output.

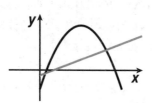

If we know the expressions that define these functions, we can use our knowledge of quadratic equations to answer questions such as:

- Will the two functions ever have the same value?

  (Yes. We can see that their graphs intersect in a couple of places.)

- If so, at what input values does that happen? What are the output values they have in common?

  (To find out, we can write and solve this equation: $f(x) = g(x)$.)

Glossary

absolute value

NAME _____ DATE _____ PERIOD _____

Practice

Using Quadratic Equations to Model Situations and Solve Problems

1. The function h represents the height of an object t seconds after it is launched into the air. The function is defined by $h(t) = -5t^2 + 20t + 18$. Height is measured in meters.

   Answer each question without graphing. Explain or show your reasoning.

   a. After how many seconds does the object reach a height of 33 meters?

   b. When does the object reach its maximum height?

   c. What is the maximum height the object reaches?

2. The graphs that represent a linear function and a quadratic function are shown here.

   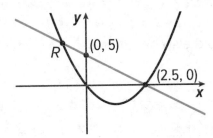

   The quadratic function is defined by $2x^2 - 5x$.

   Find the coordinates of R without using graphing technology. Show your reasoning.

3. Diego finds his neighbor's baseball in his yard, about 10 feet away from a five-foot fence. He wants to return the ball to his neighbors, so he tosses the baseball in the direction of the fence.

   Function h, defined by $h(x) = -0.078x^2 + 0.7x + 5.5$, gives the height of the ball as a function of the horizontal distance away from Diego.

   Does the ball clear the fence? Explain or show your reasoning.

4. Clare says, "I know that $\sqrt{3}$ is an irrational number because its decimal never terminates or forms a repeating pattern. I also know that $\frac{2}{9}$ is a rational number because its decimal forms a repeating pattern. But I don't know how to add or multiply these decimals, so I am not sure if $\sqrt{3} + \frac{2}{9}$ and $\sqrt{3} \cdot \frac{2}{9}$ are rational or irrational." (Lesson 7-21)

   a. Here is an argument that explains why $\sqrt{3} + \frac{2}{9}$ is irrational. Complete the missing parts of the argument.

      i. Let $x = \sqrt{3} + \frac{2}{9}$. If x were rational, then $x - \frac{2}{9}$ would also be rational because . . . .

      ii. But $x - \frac{2}{9}$ is not rational because . . . .

      iii. Since x is not rational, it must be . . . .

   b. Use the same type of argument to explain why $\sqrt{3} \cdot \frac{2}{9}$ is irrational.

NAME _____ DATE _____ PERIOD _____

5. The following expressions all define the same quadratic function.

(Lesson 7-22)

$x^2 + 2x - 8$ $(x + 4)(x - 2)$ $(x + 1)^2 - 9$

a. What is the y-intercept of the graph of the function?

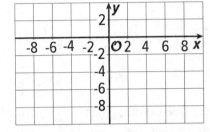

b. What are the x-intercepts of the graph?

c. What is the vertex of the graph?

d. Sketch a graph of the quadratic function without using technology. Make sure the x-intercepts, y-intercept, and vertex are plotted accurately.

6. Here are two quadratic functions: $f(x) = (x + 5)^2 + \frac{1}{2}$ and $g(x) = (x + 5)^2 + 1.$

Andre says that both f and g have a minimum value, and that the minimum value of f is less than that of g. Do you agree? Explain your reasoning.

(Lesson 7-23)

7. Function p is defined by the equation
$p(x) = (x + 10)^2 - 3$. (Lesson 7-23)

Function q is represented by this graph.

Which function has the smaller minimum?
Explain your reasoning.

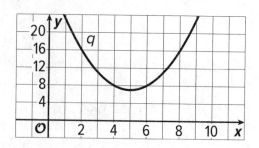

8. Without using graphing technology, sketch a graph that represents each quadratic function. Make sure the x-intercepts, y-intercept, and vertex are plotted accurately. (Lesson 7-22)

$f(x) = x^2 + 4x + 3$

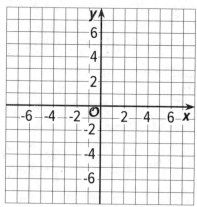

$g(x) = x^2 - 4x + 3$

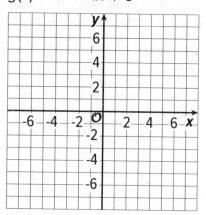

$h(x) = x^2 - 11x + 28$

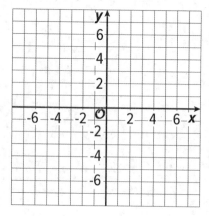

Learning Targets

Lesson	Learning Target(s)
7-1 Finding Unknown Inputs	• I can explain the meaning of a solution to an equation in terms of a situation. • I can write a quadratic equation that represents a situation.
7-2 When and Why Do We Write Quadratic Equations	• I can recognize the factored form of a quadratic expression and know when it can be useful for solving problems. • I can use a graph to find the solutions to a quadratic equation but also know its limitations.
7-3 Solving Quadratic Equations by Reasoning	• I can find solutions to quadratic equations by reasoning about the values that make the equation true. • I know that quadratic equations may have two solutions.
7-4 Solving Quadratic Equations with the Zero Product Property	• I can explain the meaning of the "zero product property." • I can find solutions to quadratic equations when one side is a product of factors and the other side is zero.

(continued on the next page)

(continued from the previous page)

Lesson	Learning Target(s)
7-5 How Many Solutions?	• I can explain why dividing by a variable to solve a quadratic equation is not a good strategy. • I know that quadratic equations can have no solutions and can explain why there are none.
7-6 Rewriting Quadratic Expressions in Factored Form (Part 1)	• I can explain how the numbers in a quadratic expression in factored form relate to the numbers in an equivalent expression in standard form. • When given quadratic expressions in factored form, I can rewrite them in standard form. • When given quadratic expressions in the form of $x^2 + bx + c$, I can rewrite them in factored form.
7-7 Rewriting Quadratic Expressions in Factored Form (Part 2)	• I can explain how the numbers and signs in a quadratic expression in factored form relate to the numbers and signs in an equivalent expression in standard form. • When given a quadratic expression given in standard form with a negative constant term, I can write an equivalent expression in factored form.
7-8 Rewriting Quadratic Expressions in Factored Form (Part 3)	• I can explain why multiplying a sum and a difference, $(x + m)(x - m)$, results in a quadratic expression with no linear term. • When given quadratic expressions in the form of $x^2 + bx + c$, I can rewrite them in factored form.

(continued on the next page)

(continued from the previous page)

Lesson	Learning Target(s)
7-9 Solving Quadratic Equations by Using Factored Form	• I can rearrange a quadratic equation to be written as expression in factored form $= 0$ and find the solutions. • I can recognize quadratic equations that have 0, 1, or 2 solutions when they are written in factored form.
7-10 Rewriting Quadratic Expressions in Factored Form (Part 4)	• I can use the factored form of a quadratic expression or a graph of a quadratic function to answer questions about a situation. • When given quadratic expressions of the form $ax^2 + bx + c$ and a is not 1, I can write equivalent expressions in factored form.
7-11 What are Perfect Squares?	• I can recognize perfect-square expressions written in different forms. • I can recognize quadratic equations that have a perfect-square expression and solve the equations.
7-12 Completing the Square (Part 1)	• I can explain what it means to "complete the square" and describe how to do it. • I can solve quadratic equations by completing the square and finding square roots.

(continued on the next page)

(continued from the previous page)

Lesson	Learning Target(s)
7-13 Completing the Square (Part 2)	• When given a quadratic equation in which the coefficient of the squared term is 1, I can solve it by completing the square.
7-14 Completing the Square (Part 3)	• I can complete the square for quadratic expressions of the form $ax^2 + bx + c$ when a is not 1 and explain the process. • I can solve quadratic equations in which the squared term coefficient is not 1 by completing the square.
7-15 Quadratic Equations with Irrational Solutions	• I can use the radical and "plus-minus" symbols to represent solutions to quadratic equations. • I know why the plus-minus symbol is used when solving quadratic equations by finding square roots.
7-16 The Quadratic Formula	• I can use the quadratic formula to solve quadratic equations. • I know some methods for solving quadratic equations can be more convenient than others.

(continued on the next page)

(continued from the previous page)

Lesson	Learning Target(s)
7-17 Applying the Quadratic Formula (Part 1)	• I can use the quadratic formula to solve an equation and interpret the solutions in terms of a situation.
7-18 Applying the Quadratic Formula (Part 2)	• I can identify common errors when using the quadratic formula. • I know some ways to tell if a number is a solution to a quadratic equation.
7-19 Deriving the Quadratic Formula	• I can explain the steps and complete some missing steps for deriving the quadratic formula. • I know how the quadratic formula is related to the process of completing the square for a quadratic equation $ax^2 + bx + c = 0$.
7-20 Rational and Irrational Solutions	• I can explain why adding a rational number and an irrational number produces an irrational number. • I can explain why multiplying a rational number (except 0) and an irrational number produces an irrational number. • I can explain why sums or products of two rational numbers are rational.

(continued on the next page)

(continued from the previous page)

Lesson	Learning Target(s)
7-21 Sums and Products of Rational and Irrational Numbers	• I can explain why adding a rational number and an irrational number produces an irrational number. • I can explain why multiplying a rational number (except 0) and an irrational number produces an irrational number. • I can explain why sums or products of two rational numbers are rational.
7-22 Rewriting Quadratic Expressions in Vertex Form	• I can identify the vertex of the graph of a quadratic function when the expression that defines it is written in vertex form. • I know the meaning of the term "vertex form" and can recognize examples of quadratic expressions written in this form. • When given a quadratic expression in standard form, I can rewrite it in vertex form.
7-23 Using Quadratic Expressions in Vertex Form to Solve Problems	• I can find the maximum or minimum of a function by writing the quadratic expression that defines it in vertex form. • When given a quadratic function in vertex form, I can explain why the vertex is a maximum or minimum.
7-24 Using Quadratic Equations to Model Situations and Solve Problems	• I can interpret information about a quadratic function given its equation or a graph. • I can rewrite quadratic functions in different but equivalent forms of my choosing and use that form to solve problems. • In situations modeled by quadratic functions, I can decide which form to use depending on the questions being asked.

Glossary

absolute value The absolute value of a number is its distance from 0 on the number line.

absolute value function The function f given by $f(x) = |x|$.

association In statistics we say that there is an association between two variables if the two variables are statistically related to each other; if the value of one of the variables can be used to estimate the value of the other.

average rate of change The average rate of change of a function f between inputs a and b is the change in the outputs divided by the change in the inputs: $\frac{f(b) - f(a)}{b - a}$. It is the slope of the line joining $(a, f(a))$ and $(b, f(b))$ on the graph.

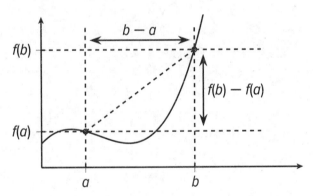

bell-shaped distribution A distribution whose dot plot or histogram takes the form of a bell with most of the data clustered near the center and fewer points farther from the center.

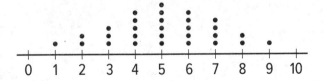

bimodal distribution A distribution with two very common data values seen in a dot plot or histogram as distinct peaks. In the dot plot shown, the two common data values are 2 and 7.

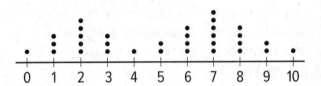

categorical data Categorical data are data where the values are categories. For example, the breeds of 10 different dogs are categorical data. Another example is the colors of 100 different flowers.

categorical variable A variable that takes on values which can be divided into groups or categories. For example, color is a categorical variable which can take on the values, red, blue, green, etc.

causal relationship A causal relationship is one in which a change in one of the variables causes a change in the other variable.

coefficient In an algebraic expression, the coefficient of a variable is the constant the variable is multiplied by. If the variable appears by itself then it is regarded as being multiplied by 1 and the coefficient is 1.

The coefficient of x in the expression $3x + 2$ is 3. The coefficient of p in the expression $5 + p$ is 1.

completing the square Completing the square in a quadratic expression means transforming it into the form $a(x + p)^2 - q$, where a, p, and q are constants.

Completing the square in a quadratic equation means transforming into the form $a(x + p)^2 = q$.

constant term In an expression like $5x + 2$ the number 2 is called the constant term because it doesn't change when x changes. In the expression $5x - 8$ the constant term is -8, because we think of the expression as $5x + (-8)$. In the expression $12x - 4$ the constant term is -4.

constraint A limitation on the possible values of variables in a model, often expressed by an equation or inequality or by specifying that the value must be an integer. For example, distance above the ground d, in meters, might be constrained to be non-negative, expressed by $d \geq 0$.

correlation coefficient A number between -1 and 1 that describes the strength and direction of a linear association between two numerical variables. The sign of the correlation coefficient is the same as the sign of the slope of the best fit line. The closer the correlation coefficient is to 0, the weaker the linear relationship. When the correlation coefficient is closer to 1 or -1, the linear model fits the data better. The first figure shows a correlation coefficient which is close to 1, the second a correlation coefficient which is positive but closer to 0, and the third a correlation coefficient which is close to -1.

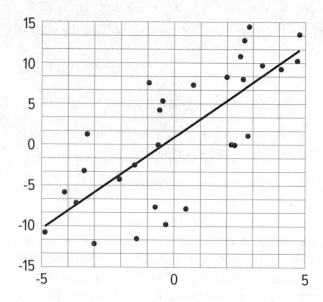

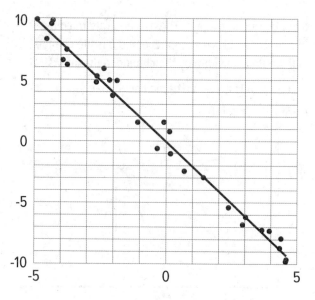

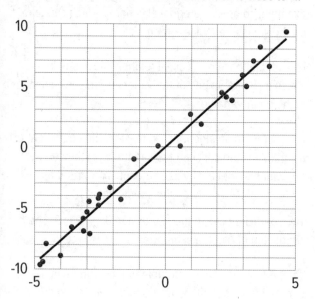

D

decreasing (function) A function is decreasing if its outputs get smaller as the inputs get larger, resulting in a downward sloping graph as you move from left to right.

A function can also be decreasing just for a restricted range of inputs. For example the function f given by $f(x) = 3 - x^2$, whose graph is shown, is decreasing for $x \geq 0$ because the graph slopes downward to the right of the vertical axis.

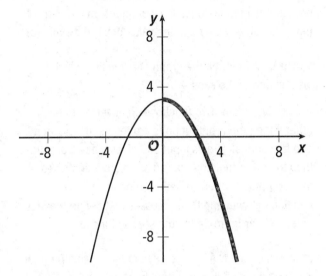

dependent variable A variable representing the output of a function.

The equation $y = 6 - x$ defines y as a function of x. The variable x is the independent variable, because you can choose any value for it. The variable y is called the dependent variable, because it depends on x. Once you have chosen a value for x, the value of y is determined.

distribution For a numerical or categorical data set, the distribution tells you how many of each value or each category there are in the data set.

domain The domain of a function is the set of all of its possible input values.

E

elimination A method of solving a system of two equations in two variables where you add or subtract a multiple of one equation to another in order to get an equation with only one of the variables (thus eliminating the other variable).

equivalent equations Equations that have the exact same solutions are equivalent equations.

equivalent systems Two systems are equivalent if they share the exact same solution set.

exponential function An exponential function is a function that has a constant growth factor. Another way to say this is that it grows by equal factors over equal intervals. For example, $f(x) = 2 \cdot 3^x$ defines an exponential function. Any time x increases by 1, $f(x)$ increases by a factor of 3.

F

factored form (of a quadratic expression) A quadratic expression that is written as the product of a constant times two linear factors is said to be in factored form. For example, $2(x - 1)(x + 3)$ and $(5x + 2)(3x - 1)$ are both in factored form.

five-number summary The five-number summary of a data set consists of the minimum, the three quartiles, and the maximum. It is often indicated by a box plot like the one shown, where the minimum is 2, the three quartiles are 4, 4.5, and 6.5, and the maximum is 9.

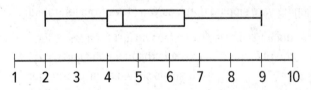

function A function takes inputs from one set and assigns them to outputs from another set, assigning exactly one output to each input.

function notation Function notation is a way of writing the outputs of a function that you have given a name to. If the function is named f and x is an input, then $f(x)$ denotes the corresponding output.

G

graph of a function The graph of a function is the set of all of its input-output pairs in the coordinate plane.

growth factor In an exponential function, the output is multiplied by the same factor every time the input increases by one. The multiplier is called the growth factor.

growth rate In an exponential function, the growth rate is the fraction or percentage of the output that gets added every time the input is increased by one. If the growth rate is 20% or 0.2, then the growth factor is 1.2.

H
---

horizontal intercept The horizontal intercept of a graph is the point where the graph crosses the horizontal axis. If the axis is labeled with the variable x, the horizontal intercept is also called the x-intercept. The horizontal intercept of the graph of $2x + 4y = 12$ is $(6,0)$.

The term is sometimes used to refer only to the x-coordinate of the point where the graph crosses the horizontal axis.

I
---

increasing (function) A function is increasing if its outputs get larger as the inputs get larger, resulting in an upward sloping graph as you move from left to right.

A function can also be increasing just for a restricted range of inputs. For example the function f given by $f(x) = 3 - x^2$, whose graph is shown, is increasing for $x \leq 0$ because the graph slopes upward to the left of the vertical axis.

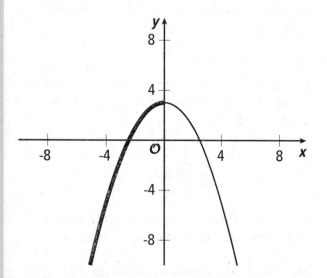

independent variable A variable representing the input of a function.

The equation $y = 6 - x$ defines y as a function of x. The variable x is the independent variable, because you can choose any value for it. The variable y is called the dependent variable, because it depends on x. Once you have chosen a value for x, the value of y is determined.

intercept A point on the graph of a function which is also on one of the axes.

inverse (function) Two functions are inverses to each other if their input-output pairs are reversed, so that if one function takes a as input and gives b as an output, then the other function takes b as an input and gives a as an output. You can sometimes find an inverse function by reversing the processes that define the first function in order to define the second function.

irrational number An irrational number is a number that is not rational. That is, it cannot be expressed as a positive or negative fraction, or zero.

L
---

linear function A linear function is a function that has a constant rate of change. Another way to say this is that it grows by equal differences over equal intervals. For example, $f(x) = 4x - 3$ defines a linear function. Any time x increases by 1, $f(x)$ increases by 4.

linear term The linear term in a quadratic expression (In standard form) $ax^2 + bx + c$, where a, b, and c are constants, is the term bx. (If the expression is not in standard form, it may need to be rewritten in standard form first.)

M
---

maximum A value of a function that is greater than or equal to all the other values, corresponding to the highest point on the graph of the function.

minimum A value of a function that is less than or equal to all the other values, corresponding to the lowest point on the graph of the function.

model A mathematical or statistical representation of a problem from science, technology, engineering, work, or everyday life, used to solve problems and make decisions.

N

negative relationship A relationship between two numerical variables is negative if an increase in the data for one variable tends to be paired with a decrease in the data for the other variable.

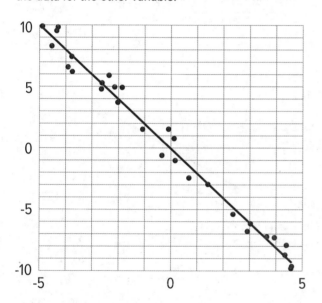

non-statistical question A non-statistical question is a question which can be answered by a specific measurement or procedure where no variability is anticipated, for example:

- How high is that building?
- If I run at 2 meters per second, how long will it take me to run 100 meters?

numerical data Numerical data, also called measurement or quantitative data, are data where the values are numbers, measurements, or quantities. For example, the weights of 10 different dogs are numerical data.

O

outlier A data value that is unusual in that it differs quite a bit from the other values in the data set. In the box plot shown, the minimum, 0, and the maximum, 44, are both outliers.

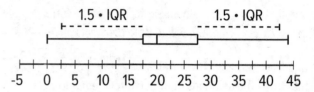

P

perfect square A perfect square is an expression that is something times itself. Usually we are interested in situations where the something is a rational number or an expression with rational coefficients.

piecewise function A piecewise function is a function defined using different expressions for different intervals in its domain.

positive relationship A relationship between two numerical variables is positive if an increase in the data for one variable tends to be paired with an increase in the data for the other variable.

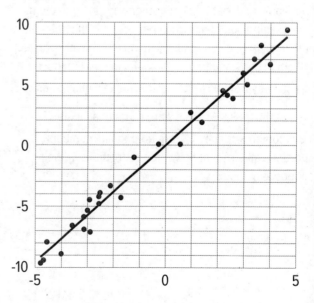

quadratic equation An equation that is equivalent to one of the form $ax^2 + bx + c = 0$, where a, b, and c are constants and $a \neq 0$.

quadratic expression A quadratic expression in x is one that is equivalent to an expression of the form $ax^2 + bx + c$, where a, b, and c are constants and $a \neq 0$.

quadratic formula The formula $x = \dfrac{-b \pm \sqrt{b^2 - 4ac}}{2a}$ that gives the solutions of the quadratic equation $ax^2 + bx + c = 0$, where a is not 0.

quadratic function A function where the output is given by a quadratic expression in the input.

R

range The range of a function is the set of all of its possible output values.

rational number A rational number is a fraction or the opposite of a fraction. Remember that a fraction is a point on the number line that you get by dividing the unit interval into b equal parts and finding the point that is a of them from 0. We can always write a fraction in the form $\dfrac{a}{b}$ where a and b are whole numbers, with b not equal to 0, but there are other ways to write them. For example, 0.7 is a fraction because it is the point on the number line you get by dividing the unit interval into 10 equal parts and finding the point that is 7 of those parts away from 0. We can also write this number as $\dfrac{7}{10}$.

The numbers 3, $-\dfrac{3}{4}$, and 6.7 are all rational numbers. The numbers π and $-\sqrt{2}$ are not rational numbers, because they cannot be written as fractions or their opposites.

relative frequency table A version of a two-way table in which the value in each cell is divided by the total number of responses in the entire table or by the total number of responses in a row or a column. The table illustrates the first type for the relationship between the condition of a textbook and its price for 120 of the books at a college bookstore.

	$10 or Less	More than $10 but Less than $30	$30 or More
new	0.025	0.075	0.225
used	0.275	0.300	0.100

residual The difference between the y-value for a point in a scatter plot and the value predicted by a linear model. The lengths of the dashed lines in the figure are the residuals for each data point.

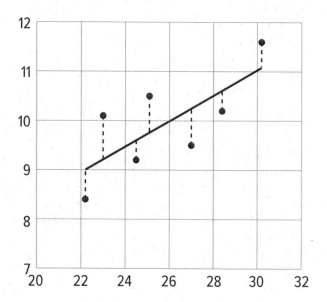

S

skewed distribution A distribution where one side of the distribution has more values farther from the bulk of the data than the other side, so that the mean is not equal to the median. In the dot plot shown, the data values on the left, such as 1, 2, and 3, are further from the bulk of the data than the data values on the right.

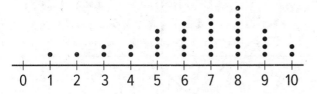

solution to a system of equations A coordinate pair that makes both equations in the system true.

On the graph shown of the equations in a system, the solution is the point where the graphs intersect.

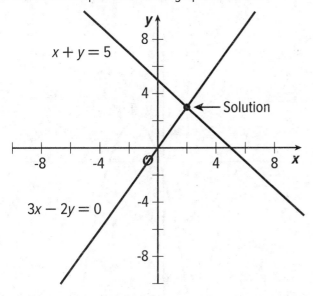

solutions to a system of inequalities All pairs of values that make the inequalities in a system true are solutions to the system. The solutions to a system of inequalities can be represented by the points in the region where the graphs of the two inequalities overlap.

standard deviation A measure of the variability, or spread, of a distribution, calculated by a method similar to the method for calculating the MAD (mean absolute deviation). The exact method is studied in more advanced courses.

standard form (of a quadratic expression) The standard form of a quadratic expression in x is $ax^2 + bx + c$, where a, b, and c are constants, and a is not 0.

statistic A quantity that is calculated from sample data, such as mean, median, or MAD (mean absolute deviation).

statistical question A statistical question is a question that can only be answered by using data and where we expect the data to have variability, for example:

- Who is the most popular musical artist at your school?
- When do students in your class typically eat dinner?
- Which classroom in your school has the most books?

strong relationship A relationship between two numerical variables is strong if the data is tightly clustered around the best fit line.

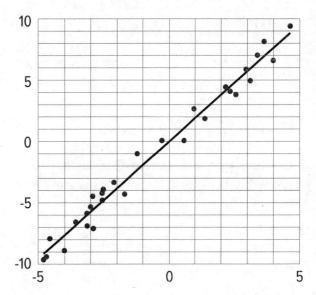

substitution Substitution is replacing a variable with an expression it is equal to.

symmetric distribution A distribution with a vertical line of symmetry in the center of the graphical representation, so that the mean is equal to the median. In the dot plot shown, the distribution is symmetric about the data value 5.

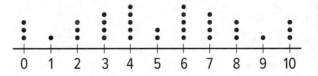

system of equations Two or more equations that represent the constraints in the same situation form a system of equations.

system of inequalities Two or more inequalities that represent the constraints in the same situation form a system of inequalities.

vertex (of a graph) The vertex of the graph of a quadratic function or of an absolute value function is the point where the graph changes from increasing to decreasing or vice versa. It is the highest or lowest point on the graph.

T

two-way table A way of organizing data from two categorical variables in order to investigate the association between them.

	has a cell phone	does not have a cell phone
10–12 years old	25	35
13–15 years old	38	12
16–18 years old	52	8

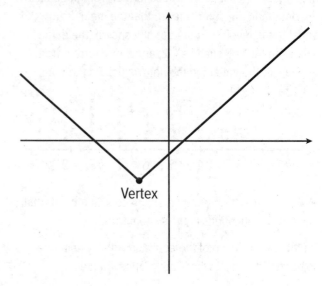

U

uniform distribution A distribution which has the data values evenly distributed throughout the range of the data.

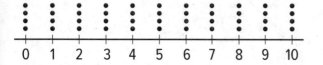

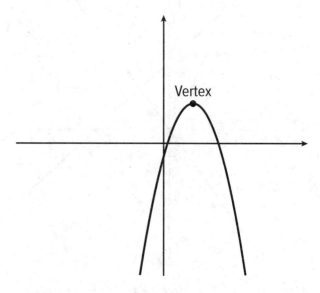

V

variable (statistics) A characteristic of individuals in a population that can take on different values

vertex form (of a quadratic expression) The vertex form of a quadratic expression in x is $a(x - h)^2 + k$, where a, h, and k are constants, and a is not 0.

vertical intercept The vertical intercept of a graph is the point where the graph crosses the vertical axis. If the axis is labeled with the variable y, the vertical intercept is also called the y-intercept.

Also, the term is sometimes used to mean just the y-coordinate of the point where the graph crosses the vertical axis. The vertical intercept of the graph of $y = 3x - 5$ is (0, -5), or just -5.

W

weak relationship A relationship between two numerical variables is weak if the data is loosely spread around the best fit line.

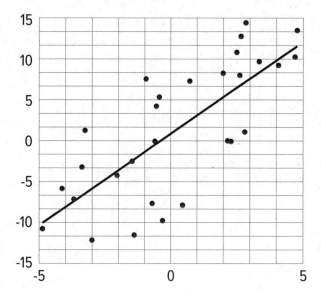

Z

zero (of a function) A zero of a function is an input that yields an output of zero. If other words, if $f(a) = 0$ then a is a zero of f.

zero product property The zero product property says that if the product of two numbers is 0, then one of the numbers must be 0.

Index